THE

BIODIESEL

SOLUTION

SAMUEL P. "PAT" BLACK, III

THE
BIODIESEL
SOLUTION

HOW BIODIESEL IS **MAKING A DIFFERENCE** FOR OUR FUTURE

Advantage®

Published by Advantage, Charleston, South Carolina.
Member of Advantage Media Group.

ADVANTAGE is a registered trademark, and the Advantage colophon is a trademark of Advantage Media Group, Inc.

Printed in the United States of America.

10 9 8 7 6 5 4 3 2 1

ISBN: 978-1-59932-849-2
LCCN: 2017962381

Book design by Megan Elger.
Photo of Terry Branstad on page 143 by Gage Skidmore.

This publication is designed to provide accurate and authoritative information in regard to the subject matter covered. It is sold with the understanding that the publisher is not engaged in rendering legal, accounting, or other professional services. If legal advice or other expert assistance is required, the services of a competent professional person should be sought.

 Advantage Media Group is proud to be a part of the Tree Neutral® program. Tree Neutral offsets the number of trees consumed in the production and printing of this book by taking proactive steps such as planting trees in direct proportion to the number of trees used to print books. To learn more about Tree Neutral, please visit **www.treeneutral.com**.

Advantage Media Group is a publisher of business, self-improvement, and professional development books. We help entrepreneurs, business leaders, and professionals share their Stories, Passion, and Knowledge to help others Learn & Grow. Do you have a manuscript or book idea that you would like us to consider for publishing? Please visit **advantagefamily.com** or call **1.866.775.1696**.

*To our everyday HEROes throughout America—HERO BX is
not a place, but a community. The use of alternative energy continues
to grow because of you all.*

You ultimately power our world.

TABLE OF CONTENTS

ABOUT THE AUTHOR

Samuel P. "Pat" Black, III founded Lake Erie Biofuels, d.b.a. HERO BX, in 2005 with a focus on creating meaningful jobs, spurring technological innovation, and producing quality, eco-friendly products. With manufacturing facilities in Erie, Pennsylvania, and Moundville, Alabama, HERO BX is now one of the largest biodiesel plants in the Northeastern United States and one of the nation's leading biodiesel producers. HERO BX bears witness to Pat Black's commitment to investing, creating, and managing companies with innovative products and capitalizing on economic opportunities to create meaningful jobs. Under his leadership he has seen to the development of over ten companies, making him a pioneer in the field of alternative energy production. He also promotes innovation and community excellence through the Black Family Foundation, supporting strategic nonprofit investments throughout Northwest Pennsylvania that leverage the fewest resources for the greatest gains in social progress. Pat Black is a life-long resident of Erie, Pennsylvania.

FUEL FOR HUMANITY

A WORD FROM THE AUTHOR

Mother Nature has a way of finishing what she sets out to do, and she has billions of years of experience in what works best. She recycles everything. She returns plants and animals—including people—to the good earth and then starts the growth all over again. Upon the old, she builds the new.

That's a fitting lesson for some of today's industrial leaders who don't think all that much about what ends up in the dump or goes down the sewer. I'm not some rabid environmentalist, but I can't stand to see a job left half done to become someone else's problem.

As a businessman, I understand the importance of seeing things through. To succeed, a business leader must be decisive and persistent. There are opportunities at every turn that must not be squandered. Whatever is wasteful should be made productive.

That is what attracted me to the biodiesel industry, which does exactly that. It turns waste to energy. It turns something unwanted into something highly valuable. In addition, it supports our farmers by giving them a new and profitable market. Farmers, in fact, were the ones who united to develop this industry, which recycles the leftovers from agricultural production into a fuel that can do so much to help our people and our planet, as you will discover in detail in the pages ahead.

In 2004, when I was looking to invest in a fuel for the future, I researched the options, and biodiesel was the clear winner. Consider this: biodiesel gives back up to five units of energy for every unit needed to make it—an efficiency far greater than petroleum, which takes more than it gives back. Consider also that a biodiesel blend of only 20 percent will reduce pollutants in emissions by half—a big plus for a world facing serious environmental issues. Biodiesel is much safer to transport as well, and is biodegradable in the event of a spill. What's not to like? I was pleased to invest in this remarkable renewable fuel!

At the same time, I wanted to invest in manufacturing to help my hometown. The city of Erie, Pennsylvania has faced a declining population as manufacturers have been bought out by multinational conglomerates and its residents have moved elsewhere to get jobs. The poverty and unemployment rates in the surrounding county are high. In a recent blow, Erie learned that GE Transportation would be ceasing its century of building locomotives in the city. I wanted to do my part to bring back more family-sustaining wages to my community, which has much to offer industries that choose to set up shop here.

If the biodiesel industry is to flourish, it needs people speaking up on its behalf and demanding that policymakers level the playing field. We must educate our elected officials and the regulators on why they must cultivate this crucial resource. The wide-ranging benefits of biodiesel are evident, and our leaders should demonstrate their support in the form of mandates, incentives, and subsidies that will encourage the industry's growth. The government has a long history of providing such support to other industries that it deems essential. Biodiesel is a relatively new player, but it has long since proved its mettle. It deserves a fighting chance to keep up the good work.

In the past decade, the federal Renewable Fuel Standard has done much to promote the growth of biofuels in this country. Moving forward, we need to pay particular attention to the role of biodiesel in meeting our national goals. Our political leaders need to get the clear message that in the years to come we will need increasingly more of this advanced biofuel. With so much at stake, we cannot afford to slip backward.

At the federal level, the Environmental Protection Agency (EPA) should be requiring greater, not lesser, amounts of biodiesel in the fuel supply; we need the congressional willpower to keep biodiesel tax credits from lapsing, and to reform them to stimulate domestic production. Historically, the tax policy has encouraged biodiesel imports at the expense of American jobs.

At the state and local levels, we need to see more mandates and incentives to promote the use of biodiesel at higher blends. A few states have adopted mandates requiring blends of 2 to 20 percent, and some offer tax credits, but many have done nothing to support biodiesel. Leaders in those states need to understand why the people care about cleaner energy, and so the people must speak up.

I believe, for example, that caring parents should demand that all school buses run on a blend of at least 20 percent biodiesel for an immediate and significant reduction in the pollutants that often waft from the exhaust pipe into the bus interior. Our elected officials also should be insisting on higher blends in the fleets of commercial transportation. Public transit, as well, should make the most of biodiesel. The upfront advantages are significant, and—since the engines require no modification to use biodiesel—the cost of switching over is nil.

We live in a diesel-powered economy. The nation's trucks, ships, trains, and planes consume countless gallons of this precious but limited resource, as do many of our heating systems, particularly in the populous Northeast. Adding biodiesel to that supply in just a 20 percent mix could reduce dependence on petroleum diesel by a fifth while greatly reducing the carbon emissions that threaten the planet. At the same time, the United States would be gaining energy security in a troubled world, where petroleum reserves often have been at the epicenter of strife. Building a stronger biodiesel industry is an easy solution that would also help our economy with badly needed jobs that pay decent wages. In communities across America, biodiesel plants have been supporting thousands of families. That's just a start toward what could be.

To that end, we at HERO BX have been doing our part to promote the continuing research for advancements in our field. Our company has established a reputation for superior quality, and we will do our utmost to maintain that distinction. We have worked with Pennsylvania State University, for example, to establish a laboratory on the campus of Penn State Erie, the Behrend College. One of the initiatives of that research lab will be finding ways to reduce the sulfur content of the raw materials used in biodiesel production to reduce pollutants even further. The lab will also be looking at ways to make biodiesel function efficiently in increasingly cold conditions.

We have invested a million dollars in test equipment at our plant to ensure we ship biodiesel that consistently meets the high standards our customers expect. Trust is everything.

Every day, I work side by side with some of the best and the brightest people in the biodiesel industry. They are here because they are passionate about what we do. Even so, we know we must groom the next generation to carry on our mission, and that is what the Behrend lab helps us to accomplish. Our industry must continue the search for breakthroughs so that biodiesel will remain unquestionably the superior choice. Research dollars must go where they will do the most good. Erie needs to find and keep the very best in local talent. Our industry, as well as our community, depend on those minds.

My roots run deep in Erie, where my father, Samuel P. Black, Jr., helped to build the Erie Insurance Company. I was a boy here. I remember my town as it was, and I dream of what it can be. After seeing a lot of places in my Navy days and in my early career, I concluded that Erie was as good as any of them, and better than most, so I came home. I looked around at what needed to be done and saw a golden opportunity.

Erie fell on hard times in the past half century. I remember when we were a thriving city with a diverse manufacturing base. At one time, we were known as the steam boiler capital of the world. A lot of those companies died off, often because the owners lacked a succession plan and eventually sold to outsiders who had no ties or loyalty to Erie and soon moved the operations elsewhere.

Nonetheless, Erie is a city with good, strong bones that is ready for its revival. Every community goes through its cycles of ups and downs, and we are in the building phase now. We have much to offer. We have four universities, with a total of about twenty thousand students, and the largest osteopathic medical school in the nation—the Lake Erie College of Osteopathic Medicine. A variety of other enterprises of various sizes are dedicated to this community. We have an eager work force with the right skills for success.

My dream is to bring industry back to town, and I am actively involved in efforts to do so. We are planning to buy outside manufacturers and transition them to Erie, where they will provide much-needed jobs. We can be strong again if we get back to our roots. Manufacturing is the basis of commerce. Most communities cannot endure on retail, service, and tourism alone. To survive, cities should be deriving at least half of their gross domestic product (GDP) from manufacturing. We need to get back to *making* stuff.

My family believes in giving back, and has long contributed to community development, social causes, and the arts in our region. The development of the biodiesel industry in our town is a significant aspect of that effort. I built our plant on a former brownfield on the shores of Lake Erie, using my own resources so that I could construct the best facility possible, free of second-guessing. Today it is a clean and thriving enterprise that provides a good living for dozens of families. We are doing the same with our plant in Alabama, a facility that we purchased out of its seventh bankruptcy. We expect soon to triple production there as we grow our business and help to build up yet another community that is happy to have us. We saw an opportunity—and we did not waste it.

At HERO BX, we are proud to be part of the story of biodiesel, which we proclaim in our company logo as the "fuel for humanity." The biodiesel industry, as you will see, has a long and fascinating history. In these pages, you will meet some of the pioneers who in the past generation have led biodiesel to new heights. You will see how the biodiesel industry today is actively addressing some of society's pressing concerns. You will learn, as well, about the struggles and challenges that the industry has faced and is facing even now.

We have come a long way, but we still have a long way to go and must not lose the momentum. We must not leave the job half done, or the next generation will inherit some big problems. Let our voices be heard. I am confident that with the support of caring citizens who recognize its potential, biodiesel will secure its rightful place as the fuel of the future.

(L R) NBB CEO Donnell Rehagen, Pat Black, & Tim Keaveney during a 2017 tour of HERO BX

SPREADING THE WORD

"I need fuel!" Doc exclaims in the final scene of the movie *Back to the Future* as he tosses a banana peel and stale beer into the power generator of his tricked-out DeLorean. Exhorting his young friend Marty to jump aboard for yet another time-traveling adventure, Doc pulls down his goggles and revs the engine to zoom skyward. "Where we're going," he declares, "we don't need roads!"

Though that scene is pure fantasy, it's one way to think about the potential of biodiesel fuel to power the engines of today and tomorrow. The 1985 film predates the major developments in today's biodiesel industry, yet that scene suggests that a fuel derived essentially from stuff we might otherwise throw away could propel us to exciting new places—provided we get on board for the journey.

In many ways, the future is now. Biodiesel, which can be readily made from animal fats and vegetable oils that otherwise would go to waste, is here to stay. It is not some counterculture concoction or fringe cause. Biodiesel has become a mainstream fuel source that provides an array of environmental, economic, and other societal benefits, and still many people know little about it. Some are uncertain whether it can be used safely and reliably as a substitute for petroleum diesel. Because they misunderstand it, they often malign it. They are unaware of the facts, or believe distortions that they may have heard.

The lack of awareness is hardly surprising, since the industry didn't exist a few decades ago. Biodiesel's potential did attract some attention during the series of oil crises in the 1970s, spawned by Middle East tensions, but interest waned after crude prices moderated. Further development would not commence for two decades. Much of the renewed interest would be motivated by agricultural concerns—in particular, how to find a market for a huge surplus of soybean oil that was a drag on farm profits. Biodiesel was the answer and, as it turned out, it offered a variety of other solutions.

It is an industry so young that many of those who launched it are still deeply involved in it. The concept, though, is nothing new. In the early 1900s, Rudolf Diesel and Henry Ford envisioned a world where the fuel of choice came from agricultural products instead of petroleum—from carbohydrates, not hydrocarbons. It is an old idea with a new urgency, and, in that sense, the biodiesel industry is indeed taking us "back to the future."

A collaboration of visionaries—farmers, scientists, entrepreneurs, and conservationists—gave birth to a movement that has worked hard to prove itself. A quarter century ago, biodiesel still was being mixed in fifty-five-gallon drums in Midwest labs. In the chapters ahead, we will meet some of the pioneers who have been there since the beginning. They are passionate people dedicated to the cause of producing and promoting a clean, safe, efficient, sustainable, home-grown, and profitable fuel. In effect, they invented an industry.

This book is specifically about biodiesel. It is not about other renewable energies such as solar, wind, and water power. Nor is it about other "biofuels," an umbrella term that includes ethanol, the gasoline additive derived primarily from corn oil. Nevertheless, ethanol is an important cousin of biodiesel, and this book will chronicle the evolution of government policies and mandates that have influenced how both have developed.

Early biodiesel

Biodiesel is a renewable fuel produced around the country and worldwide from soybean and other vegetable oils, animal fats, recycled cooking oil and grease, and other raw materials, depending on the resources most readily available in a given region. Biodiesel can replace petroleum diesel and heating oil in today's engines and burners without modifying them—most often in blends of up to 5 percent (B5) or 20 percent (B20), but sometimes all the way up to pure B100 biodiesel.

Biodicsel is not raw vegetable oil as some still imagine, although raw oils *were* used in early engine experiments. To become biodiesel, the oil or fat must be processed to remove glycerin, a valuable byproduct that is widely used in drugs, foods, soaps, and cosmetics, in addition to being used for numerous industrial purposes. This process produces *methyl esters*, the chemical name for biodiesel. Today, biodiesel plants across the United States produce anywhere from half a million gallons to more than 100 million gallons a year. And, on a small scale, thrifty farmers, truckers, and others try to save money by setting up their own processing drums at home and collecting used cooking oil from local diners to make their own fuel.

Biodiesel holds great potential as a solution to challenges on many fronts in today's society:

- Biodiesel offers environmental benefits, burning much cleaner than fossil diesel, reducing greenhouse gas emissions and particulates, and helping to meet clean-air standards. Though the newer diesel engines have been designed to operate much cleaner than the older models that are still in use, biodiesel nonetheless offers a range of environmental advantages. It is nontoxic and biodegradable in the event of a spill.

- Biodiesel makes good and efficient use of agricultural byproducts and waste that might otherwise go to landfills. Though biodiesel is agricultural in origin, its production does not subtract from our food supply or encroach on our valuable land resources. To the contrary, it helps to make the most of our farmland.

- Blended with petrodiesel, biodiesel goes a long way toward extending the critical US fuel reserves. In a politically unstable world, it offers a domestic solution to power our infrastructure. In the 1990s, after the first Gulf War, policymakers focused anew on energy security issues as well as on environmental initiatives. Both served to encourage the growth of the biodiesel industry that was emerging amid the soybean fields of middle America.

- The industry puts Americans to work. The biodiesel market in 2016 supported roughly sixty-four thousand jobs, and that figure is sure to grow if federal and state policies continue to support domestic production. That's billions of dollars flowing into the household incomes of the taxpaying citizens who are producing, handling, distributing, selling, and servicing biodiesel.

The chapters ahead will explore the past, present, and future of this fuel that has so much to offer our society—if we let it. Biodiesel advocates need to raise public awareness of its benefits, and our nation needs to muster the political will to encourage its domestic production.

It comes down to this: our society depends upon the petrodiesel that fires our engines, and, as heating oil, keeps us cozy. It's hard to imagine living without it. However, petroleum is a limited resource. There is only so much of it within the earth's crust, and yet we have a growing and seemingly insatiable thirst for it. That presents issues that biodiesel can help to solve. In other words: we need diesel; therefore, we also need biodiesel.

Because biodiesel undoubtedly will play a crucial role in meeting our energy needs, the industry needs to build a groundswell of public and political support. It's not "us against them." It's not biodiesel vs. the oil industry, and it's not biodiesel vs. agriculture. We are all partners with a common interest, each with a critical role, each trying to do its part to take care of the world's needs.

It starts with education. The biodiesel industry must spread the word and keep up the momentum. The public and the policymakers, the legislators and the regulators, must fully understand just what a good thing we have going here. It is the goal of this book to dispel the mystery and the myths and bring the facts to light—and, along the way, to tell a remarkable tale.

Donnell Rehagen

"The number-one thing for our industry is consumer awareness," says Donnell Rehagen, the chief executive officer of the National Biodiesel Board (NBB), the industry's trade association. In the early 2000s, Rehagen led the Missouri Department of Transportation to adopt biodiesel for its fleet. Consumers want freedom of choice, he says, and biodiesel lets them choose a domestically made product that promotes our nation's energy security. As more pickup trucks and cars become equipped with diesel engines, he says, it won't just be the big rigs pulling up to the diesel pumps. More and more people will turn to the biodiesel alternative.

Though the industry has been seeing fewer newcomers, according to Rehagen, many of the existing biodiesel plants have been significantly expanding their production capacity, bringing much-needed jobs to their communities. Nearly a hundred US biodiesel plants were operating in 2016. Much of the recent growth has been along the East and West Coasts, reflecting a growing demand in those markets.

The industry is preparing to meet the challenges—if the policymakers step to the plate to encourage and promote that growth.

Municipal fleets were among biodiesel early adopters

"We're a relationship business in many ways," Rehagen says. "We need to have a great rapport with federal and state governmental regulators and agencies." By keeping those lines of communication open, the biodiesel industry can continue its remarkable run. In the early 2000s, a few biodiesel plants concentrated in the Midwest were producing a total of only about twenty-five million gallons annually. After scores of new producers came on the scene, the output zoomed. The US Energy Information Administration reported that US biodiesel production in 2016 totaled 1.57 billion gallons, with an annual production capacity of 2.3 billion gallons.[1] That is phenom-

---

1   "Monthly Biodiesel Production Survey," US Energy Information Administration, Form EIA-22M, https://www.eia.gov/biofuels/biodiesel/production/table1.pdf.

enal, and yet it is a small portion of the on-road diesel market of 35 to 40 billion gallons. The industry set a goal to boost production within several years to 10 percent of that market.

"We have seen our demand grow by a couple hundred million gallons a year consistently over the last several years," Rehagen says. "That's big for any industry to have that kind of year after year growth. It provides that expectation and certainty that there is more demand out there than what we're currently meeting. In fact, we believe we can grow by much more than a couple hundred million gallons a year. We just need to have all of the policies in place that create the environment where our demand continues to grow."

Joe Jobe in DC circa 2000

The biodiesel industry in the United States is a success story of the American soybean farmers, "and they deserve the credit for that," says Joe Jobe, who was Rehagen's predecessor for nearly two decades at the NBB, watching it grow from just a few employees and a $400,000 budget, to bringing in $16 million a year by the end of his tenure in 2016. A Missouri farm boy, he became a fraud investigator for the state attorney general's office before joining the NBB, rising to the top post in 1999 at age twenty-nine. Back then, biodiesel was an experimental fuel with virtually no sales. Today, it is federally designated as an advanced biofuel and makes up 5 percent of the nation's diesel fuel supply. "Farmers tend to take a pretty long-term vision and approach, and they patiently invested, and it's now paid off for them significantly," Jobe says. He recalls the genesis of the industry soon after the first Gulf War, when "we saw Saddam light up all of those Kuwaiti oil wells and watched them burn—black smoke in the desert, all summer long—and we had a recession because of high oil prices." By developing the biodiesel industry, the farmers not only found a market for their surplus soy oil, but they also made strides toward American energy security.

In the fields, in the laboratories, and in the offices of businesses large and small across the nation, "this industry is full of passionate, true believers," says Jobe, who now runs Rock House Advisors as a consultant on business strategy and policy issues for a broad range of companies, many of which are in the biodiesel and related industries. "The highs have been very high, and the lows very low. It's been a real roller coaster," he says, citing the numerous technical and political challenges involved in creating an industry.

Biodiesel represents the spirit of American ingenuity and innovation, says Doug Whitehead, the NBB's chief operating officer who studied architecture in college but has spent much of his career building something else. The industry is proud to have created something valuable from waste, he says—but he prefers to think of it another way: "Soybean oil isn't waste. Animal fat isn't waste. Used cooking oil isn't waste. Those were products that didn't have a home"— until the biodiesel industry found them a home and cleaned them up and put them to service. "We have a great story to tell," he says.

> **Biodiesel represents the spirit of American ingenuity and innovation.**

We must tell that story. We need to do what we can to further the steady growth that the industry has experienced as people have invested their money and their dreams to build a better energy future.

Biodiesel holds the promise to help solve so many pressing challenges that this nation faces—energy issues, environmental and conservation issues, sociopolitical issues, agricultural issues, and more. In short, biodiesel is the fuel for humanity.

CHAPTER 1

COMING FULL CIRCLE

At the Paris World Exposition of 1900, fifty million visitors marveled at such innovations as the escalator, films with sound, a magnetic voice recorder, a Ferris wheel, and a remarkable new kind of engine that ignited its fuel by the heat of compression. It was Rudolf Diesel's contribution to the technological advances of the day, and it chugged away on peanut oil.

The German engineer's entry won the exposition's grand prize. His engine would become a dominant technology worldwide in the ensuing decades. It would power factories and mines, ships and submarines, locomotives, tractors, and trucks. It became the workhorse of industry and commerce.

Diesel did not specifically design his invention for vegetable oil, although the early models could run on a variety of fuels, such as kerosene and coal dust. The French government, however, requested that the world's fair demonstration use oil specifically from peanuts, which were plentiful in the nation's African colonies. France hoped the colonies could find a domestic energy source so they would not need to import coal or kerosene.

Diesel was impressed by how well his invention operated on this alternative fuel. "It worked so smoothly," he reminisced in a presentation he gave in 1912, shortly before his death, "that only very few people were aware of it." Though it was built to use mineral oil, he pointed out, the engine "worked on vegetable oil without any alterations."

Rudolph Diesel

Traveling extensively to promote his engine, Diesel also began further research on the feasibility of peanut power. The French had declined to pursue the idea, but he remained intrigued. Could peanut oil promote energy self-sufficiency in the tropics? He published his conclusion, along with analytic data: "This oil is almost as effective as the natural mineral oils," he wrote, "and, as it can also be used for lubricating oil, the whole work can be carried out with a single kind of oil produced directly on the spot." The colonies, he believed, could thereby be freed of dependence upon imported oil.

Diesel wondered whether the development of vegetable oils as a fuel source could help farmers everywhere compete in the industrial economy. "The diesel engine can be fed with vegetable oils," he wrote, "and would help considerably in the development of agriculture of countries which will use it. This may appear a futuristic dream, but I can predict with greater conviction that this use of the diesel engine may in the future be of great importance." He became a leading advocate of the concept, citing a growing body of evidence. Other researchers, however, found that castor oil was an excellent fuel. And they had discovered something else: animal fats such as whale blubber could be rendered into a fine engine fuel.

Those sources, if further developed, might one day become as important as coal and "tar products," Diesel speculated. In a 1912 presentation to a British society of mechanical engineers, he pointed out that at the turn of the twentieth century, when his engine attracted crowds at the Paris exhibition, fat oils were still on a level playing field with mineral oils. "The latter were not more developed," he wrote, "and yet how important they have since become."

Diesel understood the potential of the energy stored within the cells of plants and animal fats, and his vision remains remarkably relevant. His story foreshadows many of the questions that have played into the development of today's biodiesel industry: How do we attain energy security? How do we advance agriculture? How can we build our domestic reserves to reduce dependence on foreign oil? How can the oil industry and the biodiesel industry work together as partners on a level playing field?

Environmental concerns were not much on the radar in the early 1900s, although Diesel did warn of the need for conservation. Unlike the oil that we extract from the earth, he observed, the oil that we produce on farms is sustainable and renewable—and that, too, is a central consideration in today's energy policies. With remarkable foresight, he advocated an alternative to a limited resource: "Motor power can be produced from the heat of the sun," he wrote, "which is always available for agricultural purposes, even when all our natural stores of solid and liquid fuels are exhausted."

Miners of Sunlight

Fast forward more than a century and listen to Don Scott, the NBB's director of sustainability: "Plants are masters at harnessing solar energy, and they're masters at storing solar energy." That energy transfers to whatever consumes those plants, whether it be man or beast. Therefore, whether you dine on baked beans or beef brisket,

Don Scott

you are fueling your body on the power of the sun.

"We get new sunlight brought to us every day," Scott says. "It's the one natural resource that just keeps delivering." His words echo those of Rudolf Diesel so many years earlier.

Sunshine can also fuel our economy, Scott says. "Some have suggested that America has lost its edge in manufacturing, but we've not lost our edge in agriculture. We are the world's leader in agriculture," and, by releasing more of the sun's stored energy, we can get not only a bounty of food, but also a bounty of fuel.

For rural America that means badly needed jobs—with a cleaner environment as a fringe benefit—as the biodiesel industry joins agriculture for a stronger economy. By responsibly extracting essential resources from the land, he says, "we are mining sunlight."

Biodiesel, in other words, is a bright light that is leading the way toward helping to solve our world's energy needs. Those miners of sunlight are pursuing a purpose that makes sense economically, environmentally, and politically.

One Step at a Time

Yet another detail of Rudolf Diesel's story that remains relevant is his observation that his engine required no modifications to run on peanut oil. Today, biodiesel proponents continue to emphasize that the existing diesel engines need no retrofitting to run well on the available blends, and they operate well using biodiesel as a direct, "drop-in" replacement fuel.

Biodiesel will not harm engines, and virtually all manufacturers have given their nod of approval. They all accept at least a 5 percent blend, and most of them accept a 20 percent blend. The industry continues to work with the manufacturers on those standards. Technically, most modern diesel engines can handle pure B100 biodiesel, with only a few caveats that can be readily resolved—namely, pure biodiesel will gel more quickly in cold weather; it also tends to dissolve previously built-up carbon deposits that can then clog fuel filters. Those are not issues with the lower blends.

"Our industry has been very responsible in taking one step at a time," Scott says, "and setting fuel quality specifications and getting approvals so that we know that our fuel will work. There's no technical reason why we couldn't go to higher blends, but we haven't done all the work there yet." Nonetheless, he says, many people who understand the benefits of biodiesel would prefer to fill their tanks with the straight-up B100; "they get the difference it makes."

Steve Howell, the NBB's senior technical advisor, understands that difference, and he also understands the long road that the biodiesel industry has traveled to prove itself. He should know. He was there on the front lines in the early 1990s, setting up the research and technical programming to get the industry off the ground. In 1993,

Steve Howell

he and Alan Weber founded MARC-IV, a research and consulting firm, where they continue to champion farm initiatives—"helping agriculture build the bio-based frontier," as they put it.

"We did ten years worth of work before we sold hardly any biodiesel at all," Howell says. "That was the way the soybean farmers, who were putting in most of the money, wanted it to happen. They wanted the fuel to be technically sound. They wanted the industry to work cooperatively, stand on its own, and not try to force this fuel upon people."

Even so, it wasn't easy. "We've been paying for the sins of the ethanol industry ever since we started doing biodiesel," Howell says. Biodiesel proponents, he says, strive for a spirit of partnership with the petroleum industry and the engine manufacturers, a spirit that some felt was lacking with the development of ethanol.

Earlier initiatives to develop alternative fuels, such as methanol and compressed natural gas, did not go far, despite significant investments by the engine makers and fuel companies. "We were coming in after all these others," Howell says. "The cards were stacked against us." With time and patience, the spirit of partnership paid off. "This was an industry that had a lot of potential," he says, "but I think people didn't realize how much effort it really takes for a new fuel to penetrate the existing fuel business. We were persistent, if nothing else."

They were *plenty* else. Howell and his colleagues have made their case with solid data. "We designed the standards," he says, "so you can use B20 in any piece of existing equipment and it's going to work like diesel fuel or be better."

Over the years, biodiesel proponents have confronted the resistance of people convinced that this stuff will wreck their engines. It will clog the fuel system, ruin the injectors, cause massive equipment problems— Howell has heard it all. Those are false but lingering impressions that stem, he says, from the early experience of folks who tried to make a point by operating vehicles on raw vegetable oil. It bears repeating: biodiesel must be processed chemi-

> **Biodiesel must be processed chemically from vegetable oils and animal fats and meet high standards of quality. You need science to back up the enthusiasm. It's not as if you can just empty your deep fryer into your fuel tank and drive down the road.**

cally from vegetable oils and animal fats and meet high standards of quality. You need science to back up the enthusiasm. It's not as if you can just empty your deep fryer into your fuel tank and drive down the road.

As the old saying goes, the proof of the pudding is in the eating. As consumers and manufacturers and politicians continue to gain a greater understanding of biodiesel's benefits, the industry will continue its steady progress toward solving a host of societal concerns.

Closer to a Dream

It would seem we are coming full circle and finally moving closer to realizing Rudolf Diesel's vision of an agricultural source of energy, tapping into the power that the sun has set aside for us within the oils of plants and livestock. Biodiesel proponents consider him to be among the original pioneers who blazed a trail for their industry. In fact, National Biodiesel Day is celebrated on March 18, Rudolf Diesel's birthday. Though he didn't originally design his engine to run on vegetable oils and animal fats, he clearly embraced and advocated the concept.

In the early twentieth century, the petroleum industry increasingly advocated the use of a crude-oil refining byproduct as a fuel for the new compression ignition engine that Diesel developed. For years, the industry focused on extracting paraffin and kerosene for lamps and lanterns. The unwanted byproduct left behind was known as *distillate*. Now, with the advent of the diesel engine, the industry has found a new use for that waste and a way to boost profits. It emphasized that this was a petroleum-based product that worked just as well as any of the other fuels that Diesel tried, including peanut oil.

It required no modifications to the engine. Today, we hear a variation on this theme in the case that the biodiesel industry makes for itself.

The vegetable oil alternative, however, did not vanish from the scene. Interest before World War II centered on the desire for energy independence. England, France, Germany, Belgium, and Italy were among the nations looking into that prospect. India, China, Japan, and Portugal also tested and used vegetable oils. During the war years, some nations—including Argentina and Brazil—limited the export of seed oils so they could produce domestic emergency fuel supplies and keep their diesel imports to a minimum.

Except in times of high crude prices and shortages, however, the vegetable oil alternative didn't gain much traction, and newer engine designs required a thinner viscosity of petroleum diesel. If vegetable oils were to have a chance, something would need to change. That came in 1937 when a Belgian inventor, G. Chavanne, patented a process to remove the glycerin from palm oil. By the next year, a Brussels bus was operating on what today we would call biodiesel. Nonetheless, in the postwar years, with petrodiesel relatively cheap and plentiful, the incentive disappeared. The research was set aside and nearly forgotten.

The Birth of an Industry

All of this changed in 1973 with the oil embargo imposed by the Organization of Petroleum Exporting Countries (OPEC) in retaliation against US support of Israel during the Yom Kippur War. Scientists rediscovered that pure vegetable oil could power diesel engines—and also that it would damage them. Biodiesel research thus resumed, its momentum building and waning as crude prices seesawed amid continuing concerns over Middle East tensions and interruptions to supply.

Howell recalls the renewed interest of the 1970s: "There was some research done at the University of Idaho and a couple of other universities that looked at whether you can use just raw vegetable oil. The answer was that you can use it for a little while, but not very long because it cokes the injectors." Coke is the solid residue created when oil undergoes severe oxidative and thermal breakdown at extreme engine temperatures. The higher the temperature, the harder, blacker and more brittle the coke/deposit residue. "But," he continues, "if fuel prices go up to the roof and we don't have any diesel fuel, what do you do? Well, you can take this vegetable oil and turn it into something that's a lot thinner—that looks more like regular diesel fuel, and you can use that in an engine as is, and it can probably work." But then petroleum prices dropped again. "The prices for oils and fats, even though they were low, were still higher than the price for diesel fuel, because crude oil was so cheap," Howell notes.

The Europeans therefore took the lead on further biodiesel development. In 1985, an Austrian agricultural college became the site of the first plant designed specifically to produce biodiesel. The fuel has been commercially manufactured throughout Europe since 1992, with Germany being the largest producer.

The United States eventually followed suit in the early 1990s amid energy security, environmental, and agricultural concerns. The Clean Air Act amendments of 1990 and the Energy Policy Act of 1992 propelled the development of the biodiesel industry, at a time when TV viewers were tuning in, day after day, to images of black smoke billowing from Kuwaiti oil wells. Meanwhile, in the Midwest, soybean farmers were worried about another kind of oil—the vegetable oils produced when crushing the crop to make livestock meal. They had far too much of it to sell, and prices were suffering.

That convergence of incentives galvanized the development of biodiesel in America. "There wasn't any interest from the '70s clear up until the '90s," Howell says, "when the NBB started looking it over and saying, 'We've got excess soybean oil. There's a potential market here.' And so, the NBB started a new entire industry."

"We never gave up," Howell says. "We were always there. We always thought there were so many fundamental underpinnings of why this fuel makes so much sense. Whether you're a republican or democrat, whether you're an environmentalist or a business person, there are a lot of reasons why this makes sense. That's the reason why the fuel has persisted and survived and grown."

Back in college when he was studying molecules, Howell says he never imagined himself in the career he would pursue. He began by working with Procter & Gamble, but soon discovered that he was more of an entrepreneurial sort. "I wanted to do something that was going to give back to society, keep agriculture solid, help the environment, and make a better world for my kids—and hopefully make a little bit of money along the way.

"So, biodiesel just kind of fell in my lap," Howell says. "When we first started, we never thought it would get this big. Our goal was to take a little soybean oil off the market. We were thinking if we could produce thirty million gallons of biodicscl, we'd be a hero. Seventy million was our stretch goal back in 1993 and '94. I never thought we'd reach a billion gallons, and look at us now. We're shooting for four billion, and who knows how big it can get?"

On his Nebraska soybean farm one morning in the late 1970s, a young man named Kenlon Johannes read about an experiment that the American Soybean Association attempted. He was puzzled: why would anyone try to run a diesel engine on pure soy oil? The test hadn't gone well. Still, he was intrigued. *I wonder why they did that?* he asked himself at the time. *What's going on here?*

Johannes would find out soon enough. He went on to become a founder of the NBB, where he was the first chief executive officer, helping to shape the development of biodiesel in the United States. In the next chapter, we will hear more from him and other pioneers who, as Howell describes it, essentially created an industry "from scratch, literally out of nothing."

The story of biodiesel may be coming full circle, but there still are quite a few turns up ahead.

CHAPTER 2

THE STORY THAT MIGHT GET LOST

Old Brownie is getting up in years now, but this elderly ambassador for biodiesel keeps on teaching at a high school in Eureka, Missouri, passing on to the next generation the wisdom gained from long experience—in fact, 375,000 miles of it.

For more than two decades, Old Brownie—a 1992 Ford F250 pickup truck—carried the biodiesel banner far and wide. When it was fresh off the assembly line, the two-tone brown pickup drew crowds of the curious during its first outing at the

Old Brownie

Norborne, Missouri, soybean festival. What in the world, many

wondered, is biodiesel, and how can an engine run on something made from soybean oil?

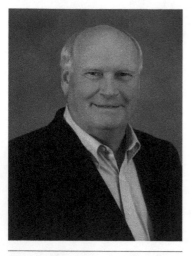

Kenlon Johannes

It was Kenlon Johannes in the driver's seat. He had been a leader in soy associations in Nebraska and Wisconsin, and at that point was executive director of the Missouri Soybean Merchandising Council. A few years earlier, Johannes reviewed a report by a pair of professors from the University of Missouri who'd studied Europe's widespread use of canola oil from rapeseeds to produce an alternative to diesel fuel. They concluded that the United States could do that, too, if only we could harvest a bumper crop of rapeseed.

Johannes set the report down. *Rapeseed?* he thought. *Why not soybeans? We have tanks full of soybean oil—why do we need another crop?* He became fascinated at the prospect of developing a domestic biodiesel industry—although, as he recalls, "we didn't even know what to call it then, *modified soybean oil*, or whatever." Perhaps, he figured, this new industry could create a demand for soy oil and soak up the glut on the market. He asked the University of Missouri to research whether soybeans could produce a good fuel, and engineering professor Leon Schumacher was soon investigating.

Where to begin? How do you kick-start an industry, particularly when few people have even heard of the product it makes? It was time to spread the word, and that is what Johannes and his colleagues set out to do. Schumacher suggested they get a truck so that people could observe how well it ran on the fuel, and their first one was a 1991 gold Dodge—"the ugliest truck on the planet," as Johannes recalls. The merchandising board could barely muster the votes to buy it.

The next question: where would they get the fuel? The truck would need to run on this new stuff, by whatever name, if it was going to impress anybody. They contacted Bill Ayres, of Interchem Inc., near Kansas City, who had been researching the process; he agreed to prepare batches derived from soybean oil to ship to Schumacher in drums. At festivals and fairs, they would "just prop up the hood and tell people it was running on this," Johannes recalls. The fuel cost several times as much as regular diesel at the time, but the investment seemed worthwhile. As interest grew, the board agreed to buy the '92 Ford the next year to further promote the concept.

Johannes and his colleagues— including Tom Verry, who would become director of outreach and development for the NBB—took that shiny new truck to county fairs and tractor pulls and trade shows and soybean grower meetings and wherever else they might get folks talking about "soy diesel."

Tom Verry

They soon found that people were talking outside the state, as well. Soybean farmers had begun a national check-off program, dedicating a percentage of their sales to fund research. "It didn't take our board members long to think, *My gosh, this is too big for just Missouri.*"

Johannes and Ayres took the truck on a promotional trip to Washington, DC, where they drove around the city and, with the help of some political connections, even managed to park for a while on the White House grounds. That got them a lot of media coverage, as well as the attention of policymakers, including Senator Tom Daschle of South Dakota, who had his photo taken with the truck, and later helped rally Capitol Hill support for biodiesel.

Meanwhile, the Missouri board got involved in another promotional trip that was dubbed the Sunrider Expedition. In 1992, Bryan Peterson of Fairfield, Iowa, contacted Johannes seeking sponsorship for a voyage around the world in his boat, a twenty-four-foot Zodiac Hurricane. He intended to operate it on biodiesel and solar power. The trip lasted over two years, with Peterson visiting dozens of countries, cities, and destinations, such as Walt Disney World, attracting considerable attention to the biodiesel cause.

Two years aboard the Zodiac Hurricane fueled by biodiesel

As the years went by, the NBB redoubled its efforts to spread the word. "It just seemed like we were flying everywhere to try and get anybody and everybody to try this thing," Johannes says. He recalls trips to every corner of the country, from Wenatchee, Washington, to the Florida Keys, from California to Pennsylvania.

The pace picked up in the mid-'90s as the NBB strove to connect biodiesel to the clean air and energy security issues that were in the national limelight. The NBB staff courted the bus companies, presenting research that a 20 percent biodiesel blend would reduce pollution effectively and inexpensively, helping them to meet regulations. It wasn't necessarily an easy sell, though. Howell recalls one of the first attempts, in 1993, when he, Ayres, and Doug Pickering paid a visit to a bus fleet manager, hoping to persuade him to give biodiesel a try. The manager kicked them out, railing that he'd had his fill of diesel additive salesmen. He was fuming more than his buses.

Still, the NBB staff made significant inroads. They hit the trade conference circuit, setting up booths at truck and bus shows, agriculture shows, fairs and expositions, mining and port shows, and more. Wherever biodiesel could help, they were there, and, slowly, the power of their message caught on. Those who tried it told others, and the momentum built.

As for Old Brownie, the veteran pickup truck took to the road for many more years, and Verry even made a tenth anniversary trip to Washington, DC in 2003. In 2016, the Missouri board donated it to the Rockwood School District, where today the biodiesel club and its teacher, Darrin Peters, feeds it on fuel that they make, in part, from the cafeteria's used cooking oil. It's a fitting retirement for a loyal stalwart of the industry.

Successes and Setbacks

The Old Brownie saga is just a footnote in the modern history of biodiesel, with its many twists and turns along the way. This is "the story that might get lost," says John Campbell, another early champion of the industry, who, in 1996, helped to build the first commercial biodiesel plant in the United States. Today he is the

John Campbell

managing director for Ocean Park Advisors, a corporate finance advisory firm with a focus on biofuels and renewable energy.

Campbell, who came from farm roots in rural Nebraska, was an animal science major in college who arrived in DC in 1979 on an internship with his congresswoman. He stayed, spending a decade promoting the ethanol industry as a solution to surplus commodities. He calls that time "the ethanol wars." He served on the staff of the senate and house agricultural committees, and from 1988 to 1991 was a deputy undersecretary of agriculture.

Then came his biodiesel days—and "there were ten or fifteen years worth of ditch digging before we got to the point that the investment climate was safe for building biodiesel plants." With his extensive experience, Campbell became a senior vice president with the soybean cooperative, Ag Processing Inc. (AGP). There he came face-to-face with the issue of the excess oil produced from crushing the beans for meal. In previous years, export subsidy and food aid programs helped with the excess, but those subsides have since been curtailed.

"The storage tanks were filling up," he says. "We, as an industry, were carrying over 3 billion pounds of soybean oil in tanks, and the soybean oil price just continued to drop. What were we going to do?" The excess was hurting the crush margins and, ultimately, soybean prices for farmers.

Meanwhile, the Missouri soybean producers took up the development of a US biodiesel industry as their mission. "Kenlon Johannes and his farmers were really the first people to get behind this on the farmers' side," Campbell says, and they were also working with entrepreneurs like Ayres and Pickering, who mixed batches of the fuel for field tests. AGP began backing them financially, helping to obtain commercial quantities of biodiesel.

A Procter & Gamble soap and oleochemical factory in Kansas City became a primary source. "We would run soybean oil into that plant," Campbell says, "and, as a side stream, they would make methyl esters, which is biodiesel, for us. And then we would ship that out for various testing—fleet tests and university tests—and keep trying to keep the pump primed with product."

In addition, Ayres and Pickering joined with AGP to investigate whether the substance could be used not just as a fuel, but perhaps as a solvent, such as paint remover, or for other applications. "We didn't know if fuel was ever going to be viable or not."

In 1992, the prospects of biodiesel finally becoming a legitimate fuel got a boost at the federal level. Campbell had maintained his White House contacts, including Robert E. Grady, an aide to President George H.W. Bush, who helped to shepherd in the Clean Air Act amendments of 1990. Working together, Campbell and Grady got a nod for biodiesel in the president's final budget. Biodiesel was to be treated in the same manner as ethanol.

Then came an unexpected setback to the research: the Procter & Gamble plant in Kansas City shut down; they'd made a corporate decision to focus on brands and marketing, abandoning biodiesel manufacturing. As a result, other factories could not produce the same quality of biodiesel from soybean oil. "If somebody didn't step up," Campbell says, "this was going to die on the vine."

AGP itself decided to take the challenge, considering that part of its mission was to add value to soybeans. Campbell says the company decided to risk a few million dollars to build a commercial-scale plant. He headed the effort, saying that it "was an investment in the future," and that "the company gave me a lot of years to nurse this along. As long as there was a light at the end of the tunnel and we kept making progress, that was enough."

The new plant opened in 1996 in Sergeant Bluff, Iowa, and within a few years other companies had built small plants in an effort to build a base for the industry. The Sergeant Bluff plant has since doubled in capacity.

"We had to prove that we had a product that would work, that wouldn't damage engines," Campbell says, "and so the soybean farmers, through their check-off dollars, were funding all of that kind of work." The industry needed federal support, including EPA approval. It needed health-effects testing. It needed approval for ASTM International technical standards and the support of engine manufacturers.

In 1998, the Asian flu epidemic was hurting agricultural commodity prices, and Campbell's contacts in Washington again proved valuable. He suggested using the power of the Commodity Credit Corporation (CCC) to purchase surplus goods. This was an opportunity for the biodiesel industry, and ultimately the Department of Agriculture established a bioenergy program under the CCC, providing a subsidy for producing biodiesel. "That program," he says, "was kind of how we survived those early years. Without it, I'm not sure if we would have made it to the point that we would be considered a legitimate fuel."

That legitimacy, however, would take about a decade longer. On Capitol Hill, biodiesel proponents heard plenty of reasons why they weren't considered ready for a blender's tax credit such as the one the ethanol industry had received. Those reasons included, at various times, a lack of ASTM International standards, a lack of approval from engine manufacturers, and a lack of commercial private-sector backing. "We could point to these plants where the private sector is investing money," Campbell says, "and we could point to the investments that the farmers had made, and tick off all the boxes to register the fuel, to get health effects testing, to get engine manufacturer approval."

Eventually, the industry won a blender's tax credit from Congress in 2005, and biodiesel was included in the Renewable Fuel Standard in 2007—"and the rest," says Campbell, "is history."

It is an industry, he says, that has always had relatively strong political and popular support. "We have never faced the opposition that ethanol faced, and continues to face," he says. "We've got plants spread all across the country, from coast to coast, and a lot of these plants use recycled cooking oils. They use low-value tallows and waste greases and all kinds of things like that, as well as soybean oil. Soybean oil is maybe 50 percent, now, of what goes into making biodiesel in the US." That diversity of interests translates into broad-based support across the country for biodiesel. "The fuel industry loves it. The truckers love it. It's just totally different from the ethanol experience."

"We Built Something Here"

All the while, the NBB played a central role in advancing the industry, using the dollars that soybean farmers contributed through the check-off program. Several state soybean associations joined forces in 1992 to establish the national group to face the challenges ahead.

High on the list was to develop this new market for the 20 percent of the bean that remained as oil after the protein was extracted for the feed market.

Originally, the groups banded together as the National Soy Diesel Development Board. A few years later they renamed it the National Biodiesel Board, in recognition of the fact that biodiesel can be produced from just about any agricultural oil or animal fats. At one point, the NBB even contacted dictionary publishers, trying to get the word "biodiesel" included in the

Soybean farmers drove the early adoption of biodiesel in the agriculture community
Dennis Wentworth, soybean farmer from Illinois

new editions—a milestone unattained until well into the new millennium, when it joined "drama queen" and "unibrow" to become an official word. It seems the publishers, like the regulators, preferred to wait until the industry proved itself. This was an instance in which a trade organization had organized on behalf of an industry that did not yet exist. That may well have been a first.

The state associations still funded most of the early work through the start of the millennium. They contributed annually to the NBB for technical work—such as the extensive research—at a cost of $2.5 million to prove to the EPA that emissions from biodiesel when burned in a diesel engine would not threaten human health.

As biodiesel plants began to dot the map in the late 1990s, the NBB served as clearinghouse and coordinator to make sure they could cross the regulatory hurdles in a simple, cost-effective way. "Some of these folks were investing millions of dollars in building biodiesel plants in the late '90s, when nobody knew anything about biodiesel," says Rehagen, appointed in 2016 to the NBB's top position. They simply used the data that the national board already had on file with the EPA, gaining quick approval to legally sell the fuel in the marketplace.

Industry leaders converge on Washington, DC in 2016 to advocate for legislation supporting biodiesel **Front** *(L-R): Alan Webber, Bob Morton, Mike Rath* **Middle***: Gene Gebolys, Ron Marr, Tim Keaveney, Senator Byron Dorgan* **Back***: Donnell Rehagen, Kent Engelbrecht, David DeRamus*

"So, the US soybean industry basically created the US biodiesel industry," Rehagen points out, "and we remain very, very strong partners to this day. I've very much been blessed to have a chance to work with the people in this industry. They continue to motivate me for the passion they bring."

Johannes, recalling his days at the helm of the NBB, is proud of what those dedicated colleagues have accomplished. "I guess I just never knew better than to think it wasn't going to work," he says. "I just said let's keep going." In effect, he says, they were staking a claim in the markets of the US petroleum industry. "I guess it never dawned on us that was what we were trying to do—and we ended up doing it."

What they did was lay the groundwork for the entrepreneurs who ultimately made it happen, "who took on the risk to invest the money to build these plants," says Johannes, a recipient of the NBB's Pioneer Award. In his modest office in 1993, Johannes recalls, he had a file cabinet in which he labeled the top drawer *Biodiesel Producers*. For several years, it was empty. It was to his great satisfaction that he finally dropped a hanging file in place there, and soon another, and then another. The industry was on its way.

Over two decades ago, biodiesel began fueling transportation

Campbell marvels at how far the industry has come in a relatively brief time. "It may not seem short, and it's been a twenty-five-year process by now. But you can look back on it and say, *Gosh. We built something here.*"

With the End Zone in Sight

Paul Nazzaro was another of those biodiesel builders. Early on, the NBB understood the need for a dialogue with the petroleum industry, and Nazzaro was the man for the job. With years of experience as a petroleum-diesel marketer in New England, he knew just what to say and do.

Paul Nazzaro

"I considered myself a painter, painting a vision," says Nazzaro, who continues to serve the NBB as its liaison with the oil industry. "With paintbrush in hand, I told a story that was pretty compelling, and it was like this: *The horizon isn't looking very good for carbon-intensive fuels. People are trying to get away from them. You know it, and I know it. It's inevitable. You need to take a look at this product.*" He told the oil companies that it was a unique fuel that they could integrate into the petroleum supply for added value to customers.

Nazzaro had made a name for himself in the petroleum industry as a marketer of gasoline, diesel, heating oil, heavy oils, and more—"virtually every line in the petroleum world," he says—and particularly as a developer of premium diesel products.

One spring day in 1996, an NBB entourage went to pitch biodiesel at Global Partners, where Nazzaro was a recent hire. "What the NBB was trying to do for their members back then was get biodiesel recognized in the transit community," says Nazzaro, and Global had the contract to sell diesel to the Massachusetts Bay Transit Authority. "Their philosophy was to send fifty-five gallon drums of biodiesel all across the country to these transits and say, 'Try it and tell us if you like it.'"

Johannes, Howell, and Weber were among the NBB visitors that day, as Nazzaro recalls. The Global executives weren't impressed with the pitch, but Nazzaro personally was intrigued. *If this fuel has the potential that they say it does,* he thought, *it could revolutionize the diesel business.*

Later that summer, he was laid off in a corporate resizing. As he agonized about what he might have done wrong, Global asked him to come back—"They loved me!"—but he would have to move to corporate headquarters in Houston. Instead, he launched Advanced Fuel Solutions, which he operates to this day, specializing in fuel-quality management. He now also has a consulting business, the Nazzaro Group.

Then he got a phone call from Howell, who'd heard that he no longer worked for Global. Howell told him that when they had met there, he clearly was the one most in tune with the biodiesel story; he was paying rapt attention, not fiddling with his papers or staring aimlessly out the window. Howell asked Nazzaro to become the NBB liaison. He asked him to take that same level of enthusiasm and open mindedness and run around the country sharing the gospel of biodiesel at refineries and pipelines and terminals and distributors and anywhere else he could make the case.

Nazzaro said he felt like a Picasso with a blank canvas, and he was never turned away, due to his reputation in the petroleum industry. He found it was an easy proposition to work with the NBB—"because what they have is what I believe in, and they value what I do for them."

"It's a unique bunch of people," he says, "and when we get together we tell stories that would make you belly-laugh."

Nazzaro delights in sharing one such tale about Howell—and it's one that Howell corroborates. More or less. The two of them were on their way to testify before a group of Minnesota legislators at a hearing on a biodiesel mandate. They stayed near the capitol at the Thunderbird Motel, where, amid the remarkable décor of animal head trophies, they changed into their business suits and headed to their meeting.

Howell looked troubled when he got out of the truck, says Nazzaro, and for good reason—the backside of his pants had ripped wide open. The only store nearby, Big Yank's, sold mostly overalls and work boots, but Howell was able to procure some black boxer shorts (less noticeable than his tighty-whities) and a sewing kit. With minutes to spare, he retreated to a men's room stall to mend his pants. "You got to do what you got to do," he said when he emerged. "This is for biodiesel, after all."

At the meeting, the stitches somehow managed to mostly hold even though Howell hadn't tied the thread off, and a long length of it trailed gracefully from his posterior as he paced at the front of the room giving his presentation. Nazzaro, within grabbing distance at the front table, says that he kept trying to pull the thread loose from his friend's pants before anyone else could see what was dangling there. Nazzaro didn't address whether the legislators seated next to him wondered what he was up to. As Howell himself recalls the scene, he kept backing up to a pillar near the podium, for fear of exposing too much. He was quite aware of the unfinished work.

Both men chuckle as they recall that day, as well as times spent side by side with many other colleagues on the biodiesel playing field. "The things we've done, the places we've been—one memory after another," says Nazzaro. "We all have a common mission—trying to get over the fifty, to the forty, and into the end zone."

This modest tale of a long-ago wardrobe malfunction serves to illustrate the persistence and the ingenuity of the biodiesel pioneers. It shows comrades with a shared mission working together to make sure that they cover all the details. It certainly confirms that they know when the end zone is in sight.

It is also one more chapter of that big story—the one that might get lost—of how biodiesel overcame the odds to become a powerhouse industry for the new millennium.

CHAPTER 3

THE NEW PIONEERS

New Englanders for generations depended on whale oil to meet their fuel needs until the beasts were increasingly scarce by the late 1800s. Meanwhile, the ascendant petroleum industry offered a cheaper alternative: kerosene could keep lanterns glowing in homes and factories and along the village streets for a third of the cost per gallon. As whalers searched the seas for a limited resource, a seemingly inexhaustible one was now gushing from the earth. Such is the age-old story of commerce and competition and emergent technologies.

Today, Tim Keaveney sees his native New England as the epicenter for another emerging fuel, which, in a way, returns us to the organic oils of yesteryear. The biodiesel industry, which produces fuel from the oils and fats of plants and animals, has made inroads to heat homes with the blend that the NBB has trademarked as Bioheat®, as it partners with the petroleum industry. Keaveney, born in Boston, is vice president of sales at HERO BX.

Keaveney and his colleagues at HERO BX are proud to be among a corps of new pioneers who are taking this phenomenal fuel to a wider range of consumers to power engines and keep homes warm. These

are the innovators and the distributors and the users who are leading the way into new territories. In this chapter, we will meet some of them. Such is the youth of the biodiesel industry that often they still work side by side with the "early" pioneers of biodiesel—the farmers and the entrepreneurs, the chemists and the environmentalists, the marketers and the investors, the politicians and the policymakers.

Tim Keaveney

Proud of his Yankee roots, Keaveney prefaces his assessment of the industry with those tales of the early whalers who once sailed from New England ports on expeditions lasting years. He recounts, for example, the sinking of the *Essex* in 1820 when a mighty sperm whale rammed it in the South Pacific and left the crew, adrift and starving, to resort to the unthinkable. The account inspired Herman Melville's novel *Moby Dick* about the vengeful quest for the "spouting fish with a horizontal tail." Keaveney is a fisherman himself, although his quest generally is for another staple of the New England waters—the native cod, prized for its oils and its delectable white fillets.

"Isn't it interesting," he says, "how the whale oil of the time that was lighting the streets of New England and New York City, in the eastern corridor, is the same place geographically where we're using much of the heating oil that's consumed in the world now. It's a bit of a *Back to the Future* story again."

Heating oil is regional. From Baltimore to Maine and westward into Ohio, "that's where heating oil happens," Keaveney says. "This unique fuel that we're making is helping fuel-oil dealers hang on to the market share they could otherwise lose to other alternative fuels, including propane and natural gas and electricity."

Keaveney cod fishing in the Gulf of Maine

Striving to develop a direct market for consumers in the region that has long been a hub for heating-oil consumption, he and his colleagues at HERO BX saw an opportunity in New Hampshire. There, "in the heart of the heating-oil world," they leased and developed a terminal for Bioheat blending and distribution in North Hampton, south of Portsmouth. "We're right on the New Hampshire seacoast, and it also serves the state of Maine, which is the coldest state in the heating-oil market," he says. "We made an investment in infrastructure to get closer to the consumer."

Two of their early customers are the longtime fuel-oil distributors, D.F. Richard Energy in Dover, New Hampshire, and Estes Oil & Propane in York, Maine. "These are old Yankee companies," says

Keaveney. "We're talking about *generations*. We're talking about *on the ocean*. We're talking about serving fishermen and lobstermen and beaches, and old colonial neighborhoods with strong ties to family."

Estes Oil and D.F. Richard have deep roots in those communities, Keaveney says, and they understand the power of solid relationships and trust with their customers. That's the recipe for their success, he says, and HERO BX shares that philosophy. His company aspires to serve not only the wholesale market, but to become a household name recognized by homeowners and the general consumer for quality, reliability, and service at a great price.

The New Hampshire terminal, Keaveney says, can store five thousand barrels of heating oil, and biodiesel can blend them electronically and systematically into "the perfect Bioheat cocktail," filling the dealers' trucks for delivery to the community. "They're looking at it as a product that's going to help them differentiate and provide the newest, cleanest emerging fuel on Earth. We want to reach the masses through them, beside them, with them."

This initiative, he says, is just the beginning of HERO BX's outreach. He says Bioheat appeals to homeowners interested in an American-made product that works better than the imports, that brings jobs to communities, and that cuts down on pollution flowing into our air and waterways. That's how forward-thinking marketers can distinguish themselves from competitors; it's not some sales pitch. The dealers truly are taking care of their customers. The homeowner or business owner gets more than just another tankful of something to make it through the winter.

Bioheat is good for the fuel dealers, and it's good for the biodiesel producers such as HERO BX. Together they are making biodiesel a fuel for all seasons. Think of a seesaw on a playground, says Keaveney. On one end is a truck driver, and on the other end is a homeowner. Imagine that playing field during the summer. The trucker teeters up. Now imagine the scene as the leaves are falling. The trucker totters down as the homeowner ascends. Since biodisel storage and handling properties are more favorable during warmer months, biodisel trucking demand peaks during March through October. In autumn—as demand slows while winter approaches—heating oil demand increases, and, since heating oil is stored and consumed indoors (where it is kept warm), there is no threat to product integrity.

In other words, the demand for over-the-road biodiesel fuel historically has dropped off after September, about the same time of year that the demand for heating oil has risen. "So, when you're a manufacturer," Keaveney explains, "and things tend to slow down right around the beginning of the fourth quarter, heating oil picks up that demand right through the winter into March and April, when demand for over-the-road biodiesel picks up again." The trucker and the homeowner take turns on the upside.

The result, over time, helps to balance biodiesel sales seasonally. The trucker is quite a big guy: over-the-road biodiesel consumption—as the big rigs pull up to the pumps at countless truck stops across the nation—is a much larger market sector than Bioheat. Still, promoting biodiesel use in heating systems not only helps the environment, but also helps to even out the year-round market.

Rick Card, the chief executive officer of D.F. Richard, knew Keaveney before his HERO BX days when he was a young emissary for the wholesale heating-oil industry. Both men are steeped in the culture of the region, and both understand its traditions and attractions. "We're an hour to Boston and an hour to the mountains," Card says, "and we're close enough but also far away enough from most everything." Anyone who lives there knows just what he means. In 1932, the company's namesake began delivering kerosene in jugs to Depression-era customers for their oil stoves. Today, tanker trucks deliver to fourteen thousand customers in a three-county region.

Pat Black with D.F. Richard Energy CEO Rick Card

The men got together again as Keaveney spearheaded the effort to get more domestically produced Bioheat into the fuel supply. Whether the consumers know it or not, their heating oil contains 5 percent biodiesel by government mandate, Card says, so they are already using it. "We should be embracing it, not hiding it," he says, pointing out that a higher blend wasn't available locally from wholesalers. HERO BX now offers a Bioheat option containing 20 percent biodiesel, and it is from a domestic source, not an offshores one.

"We have a pretty unique thing going," Card says. "We think there's nothing but a positive to gain by blending more and making it more a part of our market." HERO BX has been in the forefront of the Bioheat initiative in the region, and, he adds, "a few of us dealers have said, *Hey, this is where we're going. Let's get going.*"

Pat Black cuts ribbon to commemorate one year in business in North Hampton, NH
(L-R) John Nies, Soomi James, Tim Keaveney, Pat Black, Rick Card, Ryan Jackson, Kaleb Little

Nonetheless, his company wanted to put Bioheat cautiously to the test, monitoring its performance during that first winter. "We wanted to make sure, before really going out and singing all the praises, that it was a tried and true product—which it has been." Card warns that quality is essential. "Some dealers in the industry have started blending it themselves, what they call 'splashing' it on top, because it's cheaper. We wanted to have some science to it, and be sure that it was mixed the right way. HERO and their plant in North Hampton allows us to do that, and it gives us a quality product so that we can, in turn, be confident about what we're selling our customers."

Many homeowners think of heating oil as a commodity and will shop around for pennies of savings. "It's like gasoline nowadays," says Card. "Name brands don't matter." And yet it strikes a chord with consumers to hear they are buying a quality product made by American workers. And "we don't charge a premium for it, even though it's a premium product." In the markets where it has been available, Bioheat has been competitively priced.

Still, customers are understandably wary. *I don't want any of that stuff in my oil,* Card has sometimes heard, even though they already get 5 percent of "that stuff" in every delivery. Some have heard stories that Bioheat will damage their heating systems and gum up in cold weather. Some fuel dealers, too, wonder about such things, and Card wanted to make sure he was doing right by his customers. During that first heating season, the company began offering the blend at increasingly higher levels until, by February, with no issues reported, the deliveries to those who chose the option went to 20 percent.

As Card compared notes with other dealers who were doing the same, the consensus was that the B20 Bioheat was trouble-free. Not only that, he says, but it also helps to keep tanks clear of sludge with fewer service problems as a result. One of his drivers is also a nighttime service technician with a good perspective on exactly what he is delivering, and the results that it is producing. He reported that the blend caused not a single problem. "We're very comfortable with it," Card says, adding: "and I think the consumers will embrace it pretty quick because of all the positives to it."

Mike Estes, who serves as board chairman of the New England Fuel Institute, points out that many of his consumers already are sold on the biodiesel concept—they have just been waiting for the opportunity to put more of it in their tanks. He has been a strong supporter of renewable energy initiatives, and signed on quickly as a customer at HERO BX. Estes Oil & Propane began delivering a B15 blend and, like D.F. Richard, experienced no service issues during that first winter related to biodiesel.

An Estes Oil & Propane tanker fills up with B15 at the HERO BX blending and distribution terminal in North Hampton, NH

Most of his customers understand that climate change is a real issue, Estes says. "I decided to take a proactive approach and start putting as much blended fuel as possible into our products so that we could be delivering our customers a cleaner fuel." He and his daughter, Kate Cavanagh, operate the family business that has served southern Maine and the coastal region for well over half a century. His father, Clarence, founded it as a heating service provider in 1962. Estes took over the company in 1987 and began delivering fuel oil and, starting a decade ago, propane. Most of the customers are residential, and many are relatively affluent. The company services numerous seasonal homes while the owners are away.

"The HERO BX terminal has given us the ability to know what we're putting in a customer's tank," says Cavanagh, the company's service manager and marketing director, pointing out that the HERO BX terminal gives small dealers the opportunity to sell a product that is blended correctly to exacting standards.

Bioheat also gives customers the opportunity to burn a cleaner fuel without replacing their heating system, which can cost $7,000 to $10,000—three or four times the expense that most homeowners would expect. No changes need be made to the burners and the pumps, even though some manufacturers have not yet endorsed the use of the higher blends with their equipment. "We're mixing 15 percent, and we haven't seen any pump failures," Estes says. "It's basically seamless for us."

Because it is a cleaner product and a natural solvent, the Bioheat fuel will gradually clean out the sludge that has collected in the tank, so changing filters more often for a while is recommended. Most customers are on service plans that include those filter changes anyway, however, and no problems have been reported. "It's nothing that we've seen," Cavanagh says.

"We need to get the conversation back to the emissions, and that this is a good thing to do for the country," Estes says. "This is a good thing to do in our combat against climate change. And it's something that you can do without a huge amount of cost. You can be a pioneer in this."

Government regulations at all levels are getting increasingly tough on fossil fuels, he says, "so as things happen over the next ten to fifteen years—if we don't embrace this revolution of rebranding our product—we're going to be left at the back door without having a product to sell. This is the right thing to do for our environment."

The Pathfinders

Keaveney credits the early pioneers for clearing the path that the new pioneers are treading in the biodiesel and Bioheat industry. "You've got people like Paul Nazzaro," he says, "who's made it his life mission to try to help fuel-oil dealers capture value and hang on to market share. Through research and development, and lots of blood, sweat, and tears, they found a solution. They made this product."

Nazzaro says that, as he began working with the NBB, he quickly saw how intensely the NBB was focused on getting B20 biodiesel into the transit markets. "I said, 'I think we need to start looking at getting biodiesel recognized at the user level.'" He developed two soy-based additives with the trade names SoyShield and SoyGuard, both low-blend additives that companies could sell in bottles to improve diesel engine performance. The sales stimulated the market, he says, so that the low blends became highly recognized. "So that was phase one of, *Let's do something unique with biodiesel.*"

The next step was to get the NBB into the heating oil industry. As Nazzaro tells the story: "I'm sitting around with Steve Howell one day, and I said, 'You guys are still focused on this transit stuff. I've already shown you that we can break into markets on low blend. I've got another idea. The home-heating-oil industry is on the verge of collapse.'

"It was the heating-oil industry that I grew up in. I understood what the heating-oil industry was doing wrong. They didn't have a good product at the time. It was very dirty heat. Consumers had a very poor opinion of it." Nazzaro says that, in the late 1980s, the National Oilheat Research Alliance (NORA) surveyed consumers in its data base and found that they perceived heating oil as dirty, inefficient, and prehistoric. It was far from the ultra-low sulfur oil of today.

"I said, 'We can fix this. If this biodiesel stuff is as good as everyone says it is, and it's been very popular in the diesel market, why can't we blend it into heating oil?' Lightbulb over the head. So the name *Bioheat* was given birth in my basement in front of Steve Howell and a state oil heat executive, Michael Ferrante from Massachusetts."

Nazzaro coined the term, and Howell, wisely foreseeing the day when dealers would market the product at different blends, suggested separate names—one for up to 5 percent, another for up to 20 percent, and, yes, a third for up to 100 percent. "If that's what you want to do," Nazzaro told him, "you dream up the names." Howell put on his thinking cap and got creative: "Well, how about 'Plus' and 'Super Plus?'" And it was a go. The NBB registered the names.

And in that way, Nazzaro says, they began to stage a comeback for heating oil by bringing biodiesel into the pool—"and the reality is that it all started with *what's next?*" The residential heating oil market is only about four billion gallons a year these days, he says, down from 10.5 billion gallons, but, even so, that would be a billion-dollar market for biodiesel if all the fuel were a 20 percent blend.

"So, the East Coast is a new bastion of opportunity for biodiesel producers, clearly. You can see what they've done at HERO BX," Nazzaro says. As he pursued the management's vision, Keaveney found a local terminal. "He's taking it closer to the customer." Meanwhile, NORA has been pushing the Bioheat agenda in a move toward supporting higher blends, using money from its checkoff program for technical developments and consumer education. The time has come, Nazzaro says, for the petroleum industry to develop its own initiatives for marketing this product that gives it something to offer customers other than just another tank of fuel.

> **The time has come for the petroleum industry to develop its own initiatives for marketing this product that gives it something to offer customers other than just another tank of fuel.**

"It's all about building a future for heating oil," he adds. "Bioheat is a work in progress, and it has spawned a lot of excitement in the biodiesel community." Though B5 is the common blend, the B20 blend is becoming prevalent in some markets—such as in New York City—and, in due course, the seacoast of New Hampshire and Maine. Bioheat is the first significant development that the heating-oil industry has seen in decades, and, he says, "that's what makes it exciting, and that's why I sit back now to see: Did I do my job? Am I going to see them nurture it, or are they just going to forget? Time will tell."

Nazzaro has heard plenty of the naysaying. "'Every time there's a system issue, it's because of biodiesel.' I'm intimately involved, under my contract, with addressing every single whining dealer that loses a pump and automatically points his or her finger to biodiesel." His team's analysis, he says, clearly shows that the problems have no boundaries. They are systemwide and unrelated to whether the fuel has a biodiesel blend or additives. The problems arise from wear and tear on the pumps. "When you're the new fuel," he says, "even if you're only 5 percent of the new fuel, you're the troublemaker."

Overall, says Nazzaro, it's been a good run. Lately, his focus has turned to the West Coast, where the low-carbon fuel standard has opened up another market for biodiesel, this one for an estimated eight hundred million gallons a year. "So, between the two coasts, you have almost two billion gallons of demand," he says. "It's never a dull moment at NBB. We always have something to work on."

Jessica Robinson

Jessica Robinson, the NBB Director of Communications, calls Nazzaro the "granddaddy" of Bioheat. The integration of biodiesel into heating oil, she says, has helped suppliers remain attractive to their customers and competitive with a clean and high-performing product. "It's a cool partnership," she says, "and it's to the point where instead of the biodiesel industry trying to persuade the heating-oil market that, 'Hey, you need us,' the market is coming to the industry and saying, 'Please, we need your product.'"

Spreading the News

As a prime example of the Bioheat's success, Robinson points to the New York City market. The New York City Council recently passed legislation to require a 5 percent biodiesel blend in heating oil, and to increase that blend in increments to 20 percent by 2034. Just the increase to the B5 blend produces the environmental equivalent of taking forty-five thousand cars off the road, city officials estimated after passing the bill in September 2016, and the B20 blend will be like taking one hundred and seventy-five thousand cars off the road. They said it would decrease carbon emissions by up to 40 percent and reduce annual consumption of petroleum citywide by over 150 million gallons.[2]

The architect of the initiative is City Councilman Costa Constantinides. He says Bioheat will bring the city closer to its goal of an 80 percent reduction in carbon emissions by 2050. In addition, New York City's use of biodiesel extends well beyond Bioheat. The city's garbage trucks, snowplows, street sweepers, and other equipment all use biodiesel.

"That's a great, iconic story of America," Robinson says. "New Yorkers are not mild about how they feel about things. They love it or hate it, and biodiesel is in the love category. If somebody says, 'Oh, that can't be done,' New York will say, 'Yeah, right,' and make it happen."

For example, she says, some people fret that biodiesel doesn't cope well with cold weather. Tell that to the operators of those gigantic

2   National Biodiesel Board, "New York Governor Signs Bill Requiring Biodiesel in Heating Oil," *Biodiesel Magazine* (January 13, 2017): http://www.biodieselmagazine.com/articles/2516143/new-york-governor-signs-bill-requiring-biodiesel-in-heating-oil.

snow movers, running reliably on biodiesel, that keep the John F. Kennedy International Airport open in the face of a winter storm.

In chapter 8, Constantinides will have more to say about the New York scene and why he and other biodiesel and Bioheat advocates have been eager to start spreading the news. The message certainly has been heard at the executive mansion in Albany: Governor Andrew Cuomo, following the city's lead, signed legislation requiring at least a 5 percent biodiesel blend in all home heating oil sold in the suburban Long Island and Westchester County as of July 2018.

The Vanguard of Change

"The cool thing about biodiesel," says Robinson, "is that we have people who are into it because they're business people who see an amazing economic opportunity; we have folks who are into it because they see how much might and money it takes to maintain our oil shipping corridors, and we have passionate environmentalists looking for something else besides petroleum."

Robinson was press secretary for Matt Blunt, who, as governor of Missouri until 2009, was "a real champion for agriculture and for alternative fuel," including financial incentives for biodiesel producers and investors. Her experience made her a natural for her new job with the NBB. And there, she says, she met some of the most passionate and interesting people she'd ever encountered, and they included some of the founders of the movement. "You get to work in an industry that's so new that the people who had the ideas are still at the table, and that's pretty rare. There are very few industries like that."

Robinson has a trove of anecdotes about the biodiesel pioneers, including the early adopters who first saw the fuel's potential and showed their faith in it, and those who are taking it to new horizons today. Across the country, a growing number of entities are recognizing the value of biodiesel in helping to meet climate, clean-air, and sustainability goals. A few of Robinson's favorite stories, in no particular order, include these:

The Disneyland Railroad

To meet California's strict clean-air regulations, the Anaheim Park was considering replacing the vintage locomotives in its railroad attraction with electric engines. Instead, in 2007, the park introduced biodiesel to meet those regulations with a cleaner-burning fuel that could power the locomotives and preserve the look and feel of the steam engines of yesteryear. The resort at first used biodiesel produced from Midwest soybeans, but in 2009 it began collecting the cooking oil and grease from its hotels, restaurants, and food concessions—the waste from all those french fries and chicken nuggets—and sending it to a local biodiesel producer. Blended with a small 2 percent portion of regular diesel, the biodiesel produced from that used cooking oil met much of the locomotives' fuel requirements. On the back of postcards featuring the train attraction, which carries several million passengers each year, the park gives biodiesel its due credit for helping to keep the air clean for park visitors.

Warner Brothers Studios

Warner Brothers has used biodiesel blends since 2010 in its generators and vehicles that support the production of movies and television shows in its Burbank facilities. The studio for years has focused on conservation and renewable energy with the goal of carbon neutrality in its productions, and biodiesel has played a key role.

Florida Power & Light Co. (FPL)

Operating one of the country's largest "green fleets," this early adopter of alternative fuels deploys today about 2,500 vehicles that proudly display "Powered by Biodiesel" decals. The company recently estimated that the fleet each year, in its use of biodiesel, saves 684,000 gallons of petroleum fuel and prevents nearly 6,600 tons of carbon dioxide emissions. The drivers, who have logged more than 100 million miles on the biodiesel blends, primarily B20, have not reported issues with fuel economy or engine wear, the company says. In 2012, after the devastation of Hurricane Sandy along the East Coast, FPL crews ventured far beyond Florida to help restore power—and, with a shortage of fuel, they brought along transport trailers of biodiesel to meet not only their own requirements, but to fill the tanks of other vehicles attending to the widespread outages.

Method Products

This producer of "green" household cleaning products proudly touts its use of biodiesel, which powers the company's shipping fleet. A third of its shipments in the United States, the company says, use next-generation delivery trucks for high fuel efficiency, with the B20 blend emitting up to 20 percent less carbon and pollutants than typical trucks.

Kettle Brand Chips

Kettle Brand Chips uses sunflower and safflower oils to produce its chips, and it converts the waste into biodiesel, trademarked as "Flower

Power," to operate its fleet of vehicles. For every 7,600 bags of chips that it makes, the company estimates that it produces one gallon of waste oil—which it sends to a facility to convert into one gallon of 100 percent biodiesel. The company estimates that it prevents as much as 8 tons of carbon dioxide emissions every year.

The Planters Nutmobile

For several years, this company's peanut-shaped promotional vehicle has toured the country, employing the power of biodiesel. As part of its emphasis on sustainability, Planters made the nutmobile with reclaimed materials, including window and windshield glass and steel from old cars, and inside flooring from an 1840s barn in Lancaster County, Pennsylvania. Besides using a biodiesel blend when possible, the nutmobile also generates electricity for its interior lighting and sound system from a wind turbine and rooftop solar panels.

Great Smoky Mountains National Park

One of the nation's most visited national parks, the Great Smoky Mountains also has been establishing itself as a leader in alternative fuels and advanced vehicles. As early as 2004, it began using biodiesel in its fleet of maintenance vehicles, and today the entire fleet runs year-round on the B20 blend. The park also uses the fuel as heating oil for its headquarters' offices.

The Clark County School District

Serving the Las Vegas area and one of the largest districts in the nation, the Clark County School District was an early adopter of biodiesel. Since 2001, in an effort to provide a safer environment for its schoolchildren, the district has been using blends of B5 to B20 year-round in all of its hundreds of school buses.[3]

This is but a sampling; it is far from a comprehensive list. For every example that could be highlighted, many others have begun to do something similar as they advance the cause of biodiesel. The common denominator is passion. These are not people begrudgingly submitting to some regulation. These are the vanguards of change. They are today's pioneers who are stepping out to make the most of a fuel with the power to make a major difference in all our lives.

3   Jessica Robinson and Jennifer Weaver, "Growing Number of U.S. Diesel Vehicle Options Mirrors Growth in U.S. Biodiesel Industry," *Biodiesel.org News* (February 20, 2013): http://biodiesel.org/news/news-display/2013/02/07/las-vegas-biodiesel-vehicle-showcase-features-a-winning-combination-america's-advanced-biofuel-in-the-latest-clean-diesel-vehicles.

CHAPTER 4

A DIESEL-POWERED WORLD

The black ooze seeped for countless centuries through the rocks and into the streams of what is now northwestern Pennsylvania. Native Americans of the Erie and Seneca tribes dug pits to harvest it for medicinal purposes. The early European settlers, taking their cue from the natives, likewise began collecting this "mineral oil," believing it was good for whatever ailed them. The day would come when petroleum indeed would yield a host of pharmaceuticals, but those crude-oil curatives no doubt had side effects worse than the affliction. Soon the settlers also were using the oil as a smelly, sooty fuel for their lamps.

In 1853, George Bissell, a New York lawyer, noticed a bottle of the "Seneca Oil" remedy while visiting his alma mater, Dartmouth College, in New Hampshire. A fellow graduate, Francis Brewer, brought the sample there to show his uncle, the head of the medical school. Brewer had joined his father's lumber company in Titus-ville, Pennsylvania, where the "rock oil" seeped from springs on the property along Oil Creek. They collected it by soaking blankets in the filmy water and drying them over barrels, using the oil that was wringed out to lubricate the machinery at the mill. The oil worked as well as lard.

Bissell and his law partner, Jonathan Eveleth, were intrigued. Not only did the nation's factories need a better lubricant, but the booming population needed something cheaper and more readily available than whale oil to fill lamps. This rock oil, which was flammable as well as slippery, seemed a worthy substitute. Bissell and Eveleth recognized its immense potential to be refined into kerosene, and they joined with Brewer in 1854 to form the Pennsylvania Rock Oil Company. For $5,000, they bought two hundred acres along the Oil Creek and began their quest.

Investors, however, kept their distance, and the company's stock sold poorly. These eager entrepreneurs lacked the right technology to get the job done. They were digging pits to get to the oil, and production was slow and unimpressive. To help make their case they needed to bring in some scientific savvy, and so they hired a prominent Yale chemist to issue a report on the commercial potential for using the rock oil as both a lubricant and an illuminant. They sent him a few barrels of their oil for tests and analysis, and he sent back his conclu-sion: yes, this was good stuff from which they could extract products of significant value. They now had what they needed to stimulate interest and help give birth to an industry.

How, then, could they increase production to commercial levels to meet the growing demand for kerosene? One of their investors, a former railroad conductor named Edwin Drake, had put his entire savings of $200 into the stock and became increasingly involved in the company, which had rebranded itself as the Seneca Oil Company. Within a few years, Drake took charge of the operations in Titusville, and the company rebranded him as well, giving him the title of colonel to lend an air of prestige to the effort. It was a bit of showmanship to impress the locals.

Drake figured they could get to the oil far more efficiently by drilling for it rather than mining it. On a number of occasions, drillers in the region had struck oil instead of the drinking water or underground salt deposits that they were seeking. The oil was a disappointing nuisance,

Drake's well circa 1866

so they moved on to better ground. Drake hired an experienced salt well driller, William "Uncle Billy" Smith, and began the work, to the jeers of those who saw this as an exercise in insanity. Ignoring the naysayers, they built a derrick and purchased a steam engine to power the drill. Hitting bedrock at thirty-two feet, they continued through the summer of 1859 to slowly ram and grind their way down.

By late August, it was time to give up. The owners and investors were discouraged. Drake had burned through several thousand dollars with nothing to show for it. And by August 27, the drill had reached its maximum depth of sixty-nine and a half feet. Still nothing. Dejected, Drake and Smith pulled out the drill under orders to close up shop—and the next morning, inspecting the hole, Smith saw oil within a few inches of the top.

Soon, Drake's well was producing as much as twenty barrels of oil a day, even though the wooden derrick caught fire that autumn and the company had to rebuild. As word got out about the oil strike, competitors and fortune seekers rushed in. They built derricks all along the once quiet Oil Creek farm valley. Boom towns such as Pithole City arose in a sea of mud and bustled to a chorus of clanks, clangs, and hissing as the drills pounded incessantly, hundreds of

them. Ignorant of geology, the speculators sent down shafts at random, sometimes only yards apart.

And then it was over. Supply more than met demand in the boom, and prices plummeted. The boom towns became ghost towns. Drake's well ceased production in 1861, the company sold the property, and by 1876 the derrick was an exhibit at the Centennial Exposition in Philadelphia. Today, a replica of the well is the centerpiece of the Drake Well Museum, three miles south of Titusville in Cherrytree Township, Venango County.

Drake certainly was not the discoverer of oil, and his well in Titusville may not even have been the world's first to produce it commercially. It is historically significant though because it inaugurated the first great wave of petroleum investment in drilling, refining, and marketing. Because of Drake and his well, the business and industry had an ample supply of oil to support rampant growth—the oil boom began in Drake's well-established Pennsylvania as the hub of this new industry. The state represented half the world's petroleum production for the remainder of the century until the Texas oil boom of 1901.

Echoes From History

The more things to seem to change, the more they stay the same. In the story of the Drake well, we see a reflection of a modern story. Pennsylvania gave birth to the petroleum industry, and today it has become an important center for another developing industry: the HERO BX biodiesel plant in Erie, one of the nation's premier biodiesel producers, lies only fifty miles to the northwest of Titusville.

In partnership with petroleum, biodiesel has significantly extended our nation's energy reserves while helping to meet America's environmental objectives. Like the petroleum pioneers of the mid-nineteenth century, the biodiesel pioneers identified a high-quality replacement product that the people and the industry needed, and they capitalized on that opportunity. Like the innovators of Drake's day, they found a technology that worked, despite the jeering of naysayers. They took concerted action to make a good case for their new product, convincing the skeptics through research, analysis, and a touch of showmanship. The biodiesel pioneers, like the petroleum pioneers, took the necessary steps to attract the investment dollars needed to move forward.

Pennsylvania has remained a major player in the fossil fuels industry as well. Widespread drilling began anew a decade ago, this time in search of natural gasses trapped in the massive Marcellus shale formation that stretches across much of the state. The Marcellus Formation has since become the nation's largest source of natural gas, and has led to greater US energy security. Geologists and energy companies have long known about this buried treasure, but it runs a mile or two deep in places and they could not get to it. Technology, once again, changed everything. Advances in hydraulic fracturing— "fracking"—provided a cost-effective way to extract this resource.

Fracking has been around since the 1860s when a Civil War veteran, Colonel Edward Roberts, patented what became known as the Roberts Torpedo, dramatically increasing production in the new oil fields. The drillers hired "shooters," who filled an iron shell with gunpowder or nitroglycerin, attached a blasting cap, and lowered it into the bore hole. They then filled the shaft with water to concentrate the compression of the blast, and then dropped a weight to detonate the torpedo and shatter the rock. A trickle of oil became a torrent. It worked so well that Roberts spent a fortune on Pinkerton detectives and lawsuits as he tried to protect his patent from "moonlighters."

During the next century, the oil companies improved techniques of explosive fracking, and in the 1930s they also began using acid injections to maintain the flow. By the 1980s and 1990s, the drillers had the technology to sink a shaft vertically to great depths and then curve it horizontally to drill thousands of feet farther along the geological formation, breaking through the shale by pumping down a high-pressure mix of water, chemicals, and sand. The drillers now could go after the resources tightly trapped in the rock.

Technology provided the way. Within a decade, market forces would provide the reason. When the price was right, the drillers arrived in droves. Rural landowners cashed in. Sleepy towns became boom towns. And history demonstrated that it repeats itself.

Meanwhile, residents of those fracking communities feared for the integrity of their streams and groundwater supplies—environmental concerns are a recurring theme in the energy arena. The fracking opponents cite the risk from toxic compounds in the fracking water, and the potential that radon, mercury, lead, and arsenic could migrate upward from the fractured shale into the water as well. The environmental concerns also include greenhouse gas and air-quality issues from methane escaping the wells and from the many thousands of diesel engines operating around the clock to power the rigs, generators, and trucks. Ironically, the industry striving to meet the demand for a clean-burning fuel at a reasonable price has found itself once again associated with environmental peril.

Joining Forces

Our world needs the petroleum industry. Our infrastructure depends on it. Global oil consumption is approaching 100 million barrels a day, of which the United States has been consuming nearly 20 million barrels. Petroleum transports us, heats us, cools us. Synthetic rubber is made from it, so petroleum is in the tires as well as in the tanks of the 1.2 billion vehicles in the world (of which about a quarter are in the United States). It also yields lubricants, fertilizers, plastics, cosmetics, detergents, and nylon clothing. It provides the ingredients for medicines, cleaning fluids, and food products. We have found countless innovative uses for this essence of primordial life that the earth compressed and stored through the eons until such day as humanity could release its energy once again.

Our world also needs cleaner energy. The planet's future depends on it. Biodiesel is an important new player on the scene that can help us to meet these challenges. In antebellum America, forward thinkers developed the petroleum industry to meet the changing needs of the homeowner and the industrialist. Today, the biodiesel industry joins forces with the petroleum industry to responsibly support the infra-structure that is essential for our survival.

Diesel fuel is everywhere. It's essential for transportation, shipping and trade, and agriculture. It fuels our jets, buses, trucks, locomotives, ships, barges, and cargo-handling equipment. It powers the tractors and the combines and other equipment that cultivate our farms. It pumps the water to irrigate the fields. It fuels the trucks that get the crops and livestock to market. It moves the earth and operates the cranes for construction and infrastructure projects. It powers factories. It heats our homes, offices, schools, stores, restaurants, museums. It takes our children to and from school in all those classic yellow buses. Diesel fuel has become integral to our society. We simply can't do without it.

The United States stores an emergency oil supply of about 700 million barrels in underground caverns in Louisiana and Texas. We began building the Strategic Petroleum Reserve in 1975 after the Arab Oil Embargo interrupted the flow of petroleum from the Middle East in the aftermath of the Arab-Israeli war. US oil production at the time had been declining for several years. Although new technologies have greatly increased US oil production, reducing our dependence on foreign sources, energy security continues to be a matter of national concern. Today's inventory equates to a month or two of consumption.

The petroleum companies nowadays are plunging far deeper than the 69.5-foot shaft that struck oil in Titusville. The fracking wells now can go down 5,000 to 20,000 feet. The depth of an average Texas oil well today is about 3,500 feet. By comparison, the Grand Canyon is 2,600 feet at its deepest. The industry has gone far deeper, however, as it takes to the seas in search of this limited resource. The world's deepest well, in Russia, is 40,502 feet.

The previous record was 35,050 feet, set by the Deepwater Horizon rig in 2009 in the Gulf of Mexico. The next year, as it was drilling an exploratory well in the Gulf, an explosion sank the rig, and the blowout led to a three-month spill of four million barrels of oil, considered the worst environmental disaster of its kind in US history. Though most of the drilling is without incident, major spills have focused public opinion on the hazards rather than on the advantages of a safe, biodegradable alternative.

The growing biodiesel industry offers the potential to greatly enhance our energy independence without further risk to the environment. If the United States were to move to a standard blend of B20 biodiesel, that alone would provide a fifth of the fuel needed for our diesel engines and heating systems. That one benefit makes biodiesel an attractive investment for America. Add to that its many other advantages: it is renewable; it reduces pollutants; it is biodegradable; and, as a product derived from vegetable oils and animal fats, it is non-toxic by its nature. Biodiesel clearly is an obvious choice to become an increasingly mainstream fuel.

The Power of Repurposing

"This is the evolution of the American entrepreneurial way," says Chris Peterson, who is vice president for commodity risk and finance at HERO BX. "You go back in history to the 1800s, and innovation is the main driver. If you look at what we're doing here, we've created an industry surrounding the waste products or less desired materials from something else. Look at the petroleum industry and how it developed. Look at the Rockefeller story."

The oil refiners of the Civil War era were thinking solely about kerosene. That was their incentive to invest in an untested industry; they knew they could cash in on kerosene. The young nation's burgeoning population needed lamp fuel. Although whale oil was the best and the brightest, few could afford it anymore. A popular and inexpensive alternative was camphene, a blend of alcohol, turpentine and camphor oil. The war, however, cut off the turpentine supply from the southern pine forests, and an 1862 tax on alcohol doomed the product. Camphene is also highly volatile and could explode when lighted in a retrofitted lamp that was originally designed for whale oil. Kerosene was the answer. It burned brightly and cheaply with a relatively safe flame, and refiners could readily distill it from the crude that now was streaming from the Pennsylvania countryside.

Along with the valuable kerosene, however, came an explosively flammable byproduct called gasoline. The oil companies produced countless barrels of the stuff, and, for several decades, they disposed of it as waste. They used some of the gasoline as fuel for the refining process, but most of it they either burned off or dumped into nearby waterways, to the dismay of anyone downstream.

"We used to burn it for fuel in distilling the oil," oil tycoon John D. Rockefeller later observed, "and thousands and hundreds of thousands of barrels of it floated down the creeks and rivers, and the ground was saturated with it, in the constant effort to get rid of it." Rockefeller's Standard Oil and the other petroleum companies sold a small amount of gasoline to paint and varnish manufacturers, but they would have no real market for it until the turn of the century, when they promoted it as the perfect fuel for the rising automobile industry.

The oil companies thereby managed to turn two of their waste streams into indispensable fuels for the twentieth century. They found a huge market for the gasoline, and another huge market for the heavy distillate that they repurposed and dubbed as diesel fuel for use in Rudolf Diesel's compression engine. Likewise, the biodiesel industry of today is finding a major market for a waste stream. Peterson points out that used cooking oil was for decades destined for landfills—and "now we're making fuel out of it."

A Huge Opportunity

Clearly, since petrodiesel plays such a major role in today's world, biodiesel could also potentially play a similar role. Think about the petroleum consumed when a cargo container of plastic widgets—a device typically found in containers of beer to maintain the characteristics of the beer's head—is imported from overseas. Those widgets already are fully vested in petroleum, from which plastic is made. The widgets are delivered by diesel power to a dock, where diesel-powered equipment loads them into the container and onto a ship, which crosses the ocean under diesel power. Then the widgets are unloaded at a port—again, by diesel equipment—and reloaded onto a diesel train to be delivered by a diesel truck to their destination.

It's diesel power at virtually every step; it *could* be biodiesel power at virtually every step. Rudolf Diesel's engine has become the dominant engine of industry, commerce, and mass transit—and every one of those engines could run today on biodiesel without any modification.

"We can be proud to say that fifteen years ago, less than a quarter of the population had heard of biodiesel, and now eight out of ten people are familiar with it," says Robinson, the NBB communications director. Unfortunately, their familiarity often is quite limited, she says. They might guess that it is made from corn, for example, confusing it with ethanol. They don't connect with biodiesel because they don't connect with diesel. Few people fill up at the diesel pump.

"People don't realize," she says, "that diesel has everything to do with them. Why do I have fresh oranges in the middle of January in Missouri? Well it's certainly not because we grow them here. They got here because a truck brought them, and that truck was probably running on diesel."

"How much better would it be," she says, "if that truck was running on a 20 percent renewable fuel blend?"

Where the Rubber Meets the Road

Driving their rigs along the highways to every reach of the nation, America's truckers are by far the largest users of biodiesel. Without them, there would *be* no industry. As they pull up to the truck stops and travel centers dotting the interstate system to fill up on biodiesel blends, they're creating the demand that drives the industry. Because the nation has long depended on diesel to deliver the goods of commerce from coast to coast, today it is increasingly depending upon biodiesel as well.

The suppliers who deliver the biodiesel blends to the pumps play a crucial role in the industry's success. One of those suppliers is Lincoln Energy Solutions, the largest biodiesel reseller in the Southeast. The company has been in business for four decades, venturing into the biodiesel business in 2011. It is a major customer of HERO BX. "In the range of 15 to 20 percent of our supply comes from HERO," says Jim Farish, the company's CEO.

Jim Farish

Farish founded Lincoln in 1979 when he was only twenty-five years old. Over the years, the business branched out into terminal, transportation, and marketing companies. It was an early marketer of ethanol in the Southeast, and has been rapidly expanding its focus on delivering biodiesel to the truck stops, which can get the blends (from B5 to B20, depending on the season and location) for a penny to a nickel cheaper per gallon. Depending on the state and the blend level, the truckers may see a notice at the pump. Federal law requires a label for blends exceeding B5. Some states require the retailer to specify the level.[4]

---

4   Jon Scharingson, "Truckstop Biodiesel Pump Labeling Requirements Unraveled," NATSO.com (April 17, 2013): https://www.natso.com/blog/truckstop-biodiesel-pump-labeling-requirements-unraveled-.

Today the company serves customers in the Southeast and into the Midwest, with dozens of facilities from Jacksonville, Florida, to Tulsa, Oklahoma, and from Richmond, Virginia, to Kansas City, Missouri. Lincoln operates three petroleum terminals—in Chattanooga, Tennessee; Charlotte, North Carolina; Fredericksburg, Virginia— where it has automated equipment to blend biodiesel into the supply. Lincoln has been expanding its facilities as well, including adding a blending terminal in Kansas City, Kansas, where Farish expects demand to be increasingly strong. "Our volume has nearly doubled this year over last, and the same for 2016 over 2015," Farish says. "HERO BX has a good position in the market, and so does Lincoln."

He points out, however, that federal and state incentives, including the tax credit for blenders, are crucial to the profitability of the industry.[5] If the policymakers and the public want the environmental and economic benefit that biodiesel offers our society, they must recognize that no industry can survive without profits. The National Association of Truck Stop Operators supports biodiesel, Farish says, but "there's not enough domestic production, especially in the Southeast, to meet the demand from the truck stops." The ultimate solution is to incentivize US production while also maintaining profit margins for everyone involved in getting biodiesel into the big rigs. "More supply equals lower cost," he says.

The continued growth of the biodiesel industry will depend on the appropriate incentives to stimulate investment and demand—and that growth will depend upon greater awareness and support of biodiesel and its benefits.

---

5   "Biodiesel Tax Credit," NATSO.com (2017): https://www.natso.com/articles/sub_categories/index/biodiesel-tax-credit.

The biodiesel industry needs to cultivate an upswell of public and political support. Because much of the demand comes from the nation's truckers, they must continue filling their tanks with a product that blends together the interests of both the biodiesel and the petroleum industries. In short, that is where the rubber meets the road.

CHAPTER 5

THE BIODIESEL SOLUTION

In his hometown of Erie, Pennsylvania, Chris Peterson looks out of his office window and recalls the lower east side that he knew as a boy. It

wasn't exactly idyllic. Loads of logs rumbled in by truck and railcar to a paper mill, which for a century had spewed a stench over the neighborhood. Today, Peterson is proud to work there—although it is not the paper mill that he sees out his window. He sees a clean, high-tech facility with a splendid view of Lake Erie. The grounds are now the site of the HERO BX biodiesel plant.

Chris Peterson

Trucks and railcars still come and go, but this is a property transformed. Gone is the unpleasant smell, and gone—after diligent environmental remediation—is the pollution. Since the paper mill closed shortly after the turn of the millennium, the site has been under continuous development as a modern industrial park, bringing back much-needed jobs to the community.

Peterson was working in corporate finance in Ohio when Erie Management Group recruited him in 2006 to come home and help develop the biodiesel plant while it was still under construction. "It was an opportunity to get in on the ground floor of a startup in a relatively new industry with a lot of challenges," he says. As vice president of commodity risk and finance, he is responsible for day-to-day financial and regulatory management.

The HERO BX Erie plant

The Erie plant was "the pinnacle of technology" when it was built, Peterson says, and has remained so through a decade of investment and upgrades. The engineers in the control room are using the industry's latest equipment, says his colleague, John Nies, the vice president of operations. From their computer screens, they can open and close valves remotely and monitor the manufacturing processes. The plant is highly automated and efficient. It is more like a laboratory than a factory.

Nies was another local kid who remembers the paper mill days. As a biosystems engineering major in college, he originally pictured himself in the medical or pharmaceutical fields until he got an internship at the HERO BX biodiesel plant while it was being built, and he moved up from there. Like Peterson, he understands the facility at every level—from the characteristics and sources of the raw materials, to the

John Nies

specifications and properties of the finished fuel. In other words, these men know biodiesel in and out.

The Erie plant is one of the largest biodiesel production facilities east of the Mississippi, and among the top dozen in the nation. With a production capacity of more than 50 million gallons a year, it is one of the largest independent companies in the industry, although a few facilities owned by publicly traded conglomerates are twice or even three times as large. Some small facilities produce only a few hundred thousand gallons a year. "We have the capacity to store a little over 7 million gallons of product here on site," Peterson says, pointing to the tank farm, which is what visitors see first when entering the grounds.

VP of Operations John Nies discusses feedstock with Donnell Rehagen of the NBB at the HERO BX Erie facility

"The location has been one of the keys to our success," Peterson says. During the first year or two of operations, the company shipped over the Great Lakes and overseas to European customers while the market for biodiesel was developing domestically. Clean diesel technology has long been a mainstay in Europe, where most vehicles, including passenger cars, have diesel engines. As demand grows in the United States, particularly in the region that HERO BX serves, the company has turned from seaborne shipping to primarily truck and rail.

When the HERO BX plant was built, most of the biodiesel industry was focusing on soybean oil as the primary raw material, or "feedstock." Therefore, most new plants at the time were built close to the supply in Midwestern states such as Iowa, Illinois, and Indiana. Erie Management Group took a different tack and built close to the demand, knowing that the Northeast had a vibrant diesel market both for transportation fuel and for heating oil. The company's operations are within five hundred miles of many of the major diesel and heating oil markets in the region and in Canada.

"That closeness to our customer base has provided a significant advantage over the producers in the Midwest," Peterson says, and thus "has paid off in spades" during the facility's existence. HERO BX remains bullish on the biodiesel industry, and sees room to grow in the Northeast market, a nucleus for diesel demand.

HERO BX is also close to its major sources of raw materials that it needs to produce the biodiesel. The facility has continually invested in modifications to allow it to process multiple feedstocks and more recycled content, such as cooking oils, restaurant greases, and animal fats that once would have been destined for landfills. The company's flexibility in the raw materials that it can use has been another significant advantage.

Unlike the Midwest plants, HERO BX is not dependent on soybean oil. Instead it buys the leftovers from chicken, beef, and pork production that have been rendered into liquid fat. Much of the used cooking oil originates in restaurants, which once paid to get rid of it and now have found that they can sell it. "So, a product that historically has been truly a waste product without a lot of beneficial uses," says Peterson, "now has been transformed into a product with a lot of value."

The biodiesel industry brings jobs to communities. The Erie facility has about four dozen direct employees including plant operators, shipping and receiving staff, maintenance technicians, and the lab technicians who safeguard the company's reputation for quality. HERO BX also provides work for the truck drivers, railroad workers, and tradespeople who service the plant. "We're directly supporting forty-five or forty-six families and are indirectly helping to support several dozen more," Peterson says.

"On busy days," says Nies, "we have thirty- or forty-plus trucks in and out of here. So even though they're not our employees, we're supporting other local industries as well. We do a lot with local contractors. We try to do as much as we can with local industry."

Nies and Peterson are men who care about their hometown. They have deep roots in this city that boasts a rich maritime, railroad, and industrial heritage. They are proud of their state's abounding natural beauty and resources. As biodiesel devotees, they and their colleagues work in an industry that honors the American manufacturing tradition and respects the environment. Biodiesel plants bring jobs that pay a good living wage to communities where jobs have vanished over the decades—and the city of Erie has had its share of economic struggle. Its people have stayed in the game, positioning themselves for the rebound.

Biodiesel investment helps local economies in an environmentally conscious way. It does so in partnership with the American petroleum industry, which also has deep roots in Pennsylvania. It is the mission of HERO BX and the Erie Management Group to give back to the community, fostering pride in its history and hope for its future.

Bound for Alabama

As it surveyed the biodiesel market nationwide, HERO BX concluded that the Southeast was underserved in biodiesel production facilities and began looking for opportunities there. In 2015, the company added to its holdings a plant in Moundville, Alabama, and began developing a biodiesel customer base in that region. The supply in the Southeast was falling short of the demand, and HERO BX, like its key customer, Lincoln Energy, was alert to the possibilities for further developing its customer base.

The former owner of the Moundville plant was in bankruptcy and looking for a buyer. The facility had been a biodiesel plant for about a decade. Like many biodiesel plants, it was an old chemical plant that had been modified and repurposed. "One of the things that took the longest through the acquisition process," Peterson says, "was making sure that all of the environmental liabilities were tied up nice and neat in a bow."

Moundville is a small, rural town not too far from Tuscaloosa and the University of Alabama. The economy there has been rough for quite some time, although the region has begun to see recovery as new industries come in. The state has been aggressively courting business. Mercedes-Benz recently began a major expansion of its plant between Tuscaloosa and Birmingham, where it began producing vehicles in 1997. The manufacturing sector in the region has been growing.

The Moundville biodiesel plant provides a livelihood for dozens of employees and their families, and has a capacity of 15 million gallons a year. It is roughly a third of the size of the Erie plant. "It's not as pretty as the Erie facility," Peterson says, "and it's not as top-of-the-line or as streamlined"—at least not yet. The challenge, he says, is to "start with a clean sheet of paper and begin to do things the right way to make it work. We're making strides."

The HERO BX Moundville plant

What Goes In, What Comes Out

At both the Erie and Moundville biodiesel plants, the primary ingredients used to produce biodiesel are used cooking oil, white grease, animal fats and vegetable oils. The company has invested in the advanced pretreatment technology to deal with a wide range of raw materials—and that sets HERO BX apart from its competition.

"What that allows us to do is have flexibility in our feedstock," Nies says, explaining the significance of the technology that strips out fatty acids from animal fats and oils to produce the desired quality of biodiesel. "If used cooking oil is cheaper, we can buy more of that. When the fats market is cheaper, we can raise our blends and run more of that." Although other biodiesel plants have been striving for such flexibility, "we were ahead of the curve."

The Alabama plant does not yet have a fatty acid stripper, but it does have pretreatment capability. At both plants, technicians also put the materials through a water wash. Typically, they add a small amount of citric acid to the wash and then run it through a centrifuge to pull out metals such as sodium, potassium, magnesium, calcium, and phosphorus. Nies emphasizes that this is an environmentally friendly process. The EPA regulations are strict. "We are right on the lake, and in Alabama we're right on the Black Warrior River, and we've never had issues. In ten years, there has never been even a question that it has been completely safe."

The biodiesel process is also highly efficient. A hundred pounds of fat and oils will produce a hundred pounds of pure biodiesel fuel. Nies says that people sometimes point out that biodiesel facilities are using natural gas—and therefore, they ask, what is gained if it takes a petroleum-based fuel to produce a fuel that is not petroleum-based? The answer: much is gained. Research bears it out. For every unit of energy consumed to make biodiesel, we get 5.5 units of renewable energy—and the efficiencies keep getting better. Compare that to gasoline, which yields just 0.74 units of usable energy for every unit invested.

The feedstocks arrive at the Erie plant from a wide region. The company obtains some of it locally, and from the Buffalo area, but much of it comes from the East Coast population centers of New York and the other major cities—where a lot more people eat at a lot more restaurants, which produce a lot more cooking oils and grease. The company gets some of its animal fats from the Southeast as well as from the Chicago area. "We're strategically located," Nies says, "so that we can hit all of those markets toward the Midwest and toward the East Coast for supplying our feedstock."

How does HERO BX vet the quality of those materials? Years of research and trial and error have gone into those decisions, Nies says. "We have good relationships with a lot of our suppliers, so we know what their quality is. We trust what they're bringing in." If a supplier consistently delivers bad loads that cost too much to clean up, that supplier ceases to be a source. The company obviously also is looking for good deals on good quality. "Fat prices go up and down, so price is a huge factor. And we have a lot of data and years of research that go into knowing what blends our plant can handle and still produce the highest-quality fuel."

Joe Falk, director of supply chain management at HERO BX, talks with Donnell Rehagen of the NBB during a 2017 plant tour

In other words, this is a high-tech operation with high standards that deals with people of high standards. They're not hauling in old chicken guts. "Six, seven years ago it was much worse," Nies says.

"We've seen everything from straws to napkins to rubber gloves caught in our strainers coming in." The collectors these days do a far better job of cleaning the rendered fats and oils. Those that couldn't or wouldn't keep up with best practices went out of business. Occasionally the plant operators still see some "nasty" feedstock, he says, but he has also seen loads with "salad oil" listed on the bill of lading. It looks so pure you would think you could serve it for supper.

Once the technicians accept an incoming load, they basically clean it further until it is nearly 100 percent triglyceride fat. To that, they add methanol and a catalyst, sodium methylate. Other biodiesel facilities use sodium hydroxide, or lye, as the catalyst, but the methylate is more efficient, Nies says: "It's just more methanol going into the process." Like lye, it is caustic, but the process neutralizes it into a harmless salt.

For every hundred pounds of triglycerides, the technicians need ten pounds of methanol to initiate the reaction, and an excess ten pounds to drive it to completion. They also need a pound of the catalyst. The fluids and the catalyst go into a high-shear mixer. And what comes out is the methyl ester known as biodiesel, with glycerin as the primary byproduct. This process yields a hundred pounds of biodiesel and ten pounds of glycerin.

That's the easy part. The chemistry is simple. You could make biodiesel in a five-gallon bucket if you were so inclined. To make it efficiently and economically, though, you need to recover that 10 percent of excess methanol, and you need to purify the glycerin so you can sell it rather than pay someone to haul it away. That's where the citric wash comes in, followed by the centrifuging and thorough drying to remove traces of remaining moisture.

Glycerin (from the Greek word *glykys*, meaning *sweet*) has a multitude of industrial applications once it is refined from the crude byproduct of biodiesel distillation. Pure glycerin is colorless, odorless, nontoxic, and mildly sweet. It is suitable as a low-glycemic sweetener for diabetics, although it is not low-calorie. The biodiesel industry sells much of its glycerin for use in animal feed. Some is refined for the pharmaceutical industry, and some for cosmetics, lotions, and moisturizers. HERO BX has sold a lot of it for conversion into a product for municipal wastewater treatment, a fitting application for a company and an industry that are proud to be environmentally responsible.

The payload from the distillation process (the "transesterification") is the biodiesel. This is a one-for-one metamorphosis: into the mixer goes a pound of waste fats and oils, and out of the mixer comes a pound of biodiesel. You could fill the tank of your diesel-powered truck or car or heating system, and it would do just fine. This is what HERO BX ships from its plant. It is pure B100 biodiesel, and it

An App for That...

As the word spreads on the various advantages of biodiesel fuel, interest is growing. Trucking and bus fleets are converting to biodiesel, some at high blends. BioFuel Oasis, a worker-owned cooperative in Berkeley, California, offers a certified B99.9 blend made of recycled vegetable oil at its station where drivers can pull up to the pumps. So how can you find out where to fill your vehicle's tank with biodiesel? The NBB has an app for that—BiodieselNow. Drivers can download it at their online app store and it will locate the nearest biodiesel retailer and list the blend levels offered. Most of the diesel fuel sold in the United States already contains some biodiesel, as much as 5 percent, without being labeled. The app points out where to buy labeled biodiesel, whether it's B10, B20, or B99. The Department of Energy's Clean Cities program also offers an app—the Alternative Fueling Station Locator—to locate alternative fueling stations, including those that service propane- and electric-powered vehicles.

goes to blenders who mix it with petroleum diesel to prepare it for sale in the marketplace as B5, B20, etc. Some customers are obligated parties who must blend to meet government mandates, while others are discretionary blenders who can exceed those requirements.

The trucking industry consumes most of that product, although the demand for Bioheat is creating a major market in the Northeast. With its new terminal in New Hampshire, HERO BX is making inroads into that market as it provides blends for the heating oil suppliers, mostly at percentages up to B20. "We could theoretically go higher, but the customer so far has wanted B15 and B20 home heating oil," Nies says. "We're looking to expand that in the near future." From its production plants, however, HERO BX generally ships B100, although it does have blending capability.

A Reputation for Quality

Around the country, some biodiesel enthusiasts have set up their own stills in their garages and barns. They buy a kit and supplies, collect used cooking oil, and make homegrown fuel to operate their trucks and cars and tractors. They exhibit the same spirit of innovation and self-sufficiency that motivated the biodiesel pioneers.

"It can be a great thing," says Nies, "but if they don't really know what they're doing, it also can give the industry a bad name." This backyard biodiesel, if it is made incorrectly and falls short of standards, can clog filters and cause other troubles. "So, it's kind of a catch-22," he says. "I think it's positive for the industry to get people involved and understanding it, but if they're doing it incorrectly—and people hear the horror stories—it gives the industry a black eye."

Biodiesel manufacturers have much at stake in preserving their reputation, and HERO's brand is well known in the industry for its excellence. Missteps anywhere, however, can reflect on everyone who produces biodiesel. Anyone can make a batch. "The hard part is cleaning it, refining it, and doing it economically, Nies says. "That's where I think our company has been able to set itself apart and be successful."

"We put a massive emphasis on fuel quality," says his colleague Keaveney. "If you talk to the pioneers, people like Howell and Nazzaro, this industry had a black eye in the early days because people just didn't make high quality biodiesel." The required specifications were minimal. "The industry stumbled in its early days,"

Keaveney says, "but over the past decade the specifications for biodiesel have improved dramatically. We take that very seriously."

The industry today abides by strict regulations and standards. HERO BX is certified as a BQ-9000 producer by the National Biodiesel Accreditation Committee for producers and marketers. The program looks at storage, sampling, testing, blending, shipping, distribution, and fuel-management practices. HERO

HERO BX Technical Director Holly Koziorowski shows CEO Pat Black a sample of the company's high quality finished biodiesel

also meets the EPA standards of quality assurance in feedstocks and accounting practices. The company voluntarily submits to third-party auditing to certify compliance.

"We have outstanding employees, and outstanding customers who trust us," Keaveney says. "Trust is all we have. It sounds old-fashioned, but people do business with HERO BX because they trust us, and they trust our products. And that is the absolute truth."

Biodiesel and Petrodiesel, Side by Side

How does biodiesel compare, side by side, with petroleum diesel? Since they are partners more than they are competitors, a better question might be how well do they play together as a blend in engines and heating systems. However, we can look at some facts that researchers have established.

The NBB reports these statistics:

- Biodiesel reduces lifecycle greenhouse gases by 86 percent. It lowers particulate matter by 47 percent, reducing smog and making the air healthier to breathe. It reduces hydrocarbon emissions by 67 percent.

- Biodiesel is naturally biodegradable and nontoxic—a huge advantage in the event of a spill. Research shows that in both soil and water, biodiesel degrades four times faster than petrodiesel, with nearly 80 percent of the fuel's carbon converted by micro-organisms in as little as twenty-eight days. Biodiesel is less toxic than table salt.

- For every unit of energy that it takes to produce biodiesel, five and a half units of renewable energy are returned, the best of any US fuel.

- The National Fire Protection Association ironically classifies biodiesel fuel as nonflammable. It has low volatility and a flashpoint higher than no. 2 diesel, making it safer to store, handle, and transport. If a truck crashes, the fuel is less likely to catch fire. It does not combust until heated well above the boiling point of water.

- Biodiesel offers similar power to diesel fuel. A major advantage is that it can be used in existing engines and fuel injection equipment with little change in performance. Biodiesel has a higher cetane value, which is essentially the diesel equivalent to a higher octane rating in the gasoline world. Over millions of miles and many applications, biodiesel has proved itself similar to petrodiesel in fuel consumption, horsepower, torque, and haulage rates.

- Compared with petrodiesel, biodiesel offers significantly better lubricity, meaning it is more slippery and causes less wear and tear on engine parts. Tests have shown that even a 1 percent blend of biodiesel can improve lubricity by up to 65 percent.

- With the advent of low sulfur diesel fuel, most manufacturers have switched to engine components suitable for biodiesel. In older engines, high biodiesel blends eventually can degrade fuel lines and pump seals, and those should be upgraded. That has not been an issue with blends of B20 or lower, however. Biodiesel is more chemically active as a solvent than petrodiesel. It therefore can gradually clean out deposits in engines that have been running on petrodiesel. Filters should be checked and changed, particularly after introducing a higher blend.

- In cold weather, both petrodiesel and biodiesel can cloud and even gel. It is untrue that biodiesel becomes unmanageable in freezing weather. Generally, both require the same precautions. The lower blends are quite similar to no. 2 petrodiesel in their reaction to cold. Even a B20 blend will begin to experience cold-weather issues at only a few degrees higher than petrodiesel.

Facts are powerful, and taken together they make an irresistible case for biodiesel. It works well in existing engines and heating systems. It reduces pollutants and helps to meet emissions standards. It makes good use of waste streams. It promotes agriculture and conservation. It is safer to use. It is a renewable resource that complements a critical one that is nonrenewable. It puts Americans to work. It promotes American energy independence.

Biodiesel is part of the solution. Our society should be embracing it and advancing its use as swiftly as possible. This is not a matter of whether we should do this. We *must*.

CHAPTER 6

ENERGY FROM WASTE

When Don Scott was a boy, his family visited the historic mining town of Telluride, Colorado. There, in the box canyon at the foot of an old mule trail leading high into the Rockies, he looked up to see the glint of a waterfall in the distance. He wanted to see it up close and survey the world from that vista, but he would have to wait. The family car wasn't up to the climb.

Scott grew up to become the sustainability director for the NBB, but, in those days, he wasn't thinking biodiesel. He was thinking freedom. He was a boy dreaming of going places.

As he got older and his love of nature grew, Scott fixed up an old car so that he could travel the country—and so that he could travel to a job to earn enough to travel the country. He wanted to explore mountains and streams and caves. He wanted to do his part to protect them and make a difference in the world. But this was a practical matter. To accomplish anything, he needed money and he needed wheels.

Scott got his introduction to biodiesel during his freshman year at the University of Missouri in Columbia, where he studied civil engineering with a focus on environmental protection. "That was the same time, in 1991, that they were doing the first soy diesel development," he says. "The farmers were trying to figure out if they could use this surplus vegetable oil for fuel. And I thought that was interesting, but at the time they weren't really focused on the environmental benefits. They were really just focused on the economics. I wanted to protect the environment. I really appreciated the outdoors."

After graduation, he joined the Missouri Department of Natural Resources as an environmental engineer working to protect drinking water. It was a thirty-mile commute each way from Columbia to Jefferson City. "I had this nagging feeling that I was still doing more damage every day because of fossil fuels." He was increasingly aware that there must be a better way than to burn so much gasoline.

"To be honest, the first wave that I felt was just the economics of it," he says. "I was driving an old Jeep Grand Wagoneer that got ten miles per gallon." When gas prices spiked up, he switched to a diesel pickup to cut his fuel expenses. That's when he tried biodiesel in the tank. His first motivation was to protect his wallet, and the second was to protect the environment. His experience with renewable fuels developed into a passion, however, and in 2007 he went to work for the NBB.

Scott's journey into the world of biodiesel is much like the journey of those Missouri farmers who were working with the professors and researchers back in the early 1990s when he was just a college freshman finding his way. The farmers appreciated the outdoors, certainly, and they cared about the environment, but their foremost concern was a practical, economic matter. The first motivation for developing a biodiesel industry was to support crop prices. The farmers needed a market for the flood of surplus soy oil that is produced when soybeans are crushed and milled. What better way to develop that market than to build an industry with a host of other benefits? Good for agriculture, good for the planet—a perfect pairing.

Creating New Value

From time to time, biodiesel opponents will advance the notion that the industry diverts soybeans from the food supply. They say the soybean crop should fill hungry bellies at home and abroad and should not be squandered on filling the fuel tanks of commerce and industry. They point to the wide expanses planted in soybeans and insist those fields should instead be devoted to feeding the world, not producing biodiesel. Some blame biodiesel for a loss of wildlife habitats, and push for conservation over consumption. Some even imagine that the industry slaughters animals to harvest their fat to use as a raw material for the biodiesel plants.

Those are sadly misinformed sentiments. Here is the truth: the production of biodiesel doesn't waste food. Instead, it makes energy from waste. Biodiesel is manufactured from the largely unwanted products—the plant oils and animal fats—that are left over from the production of food.

In other words, the biodiesel industry creates value where it did not previously exist. American farmers do not grow acres of soybeans to make biodiesel. They grow the crop primarily for the protein value of the soy meal. Nor do the farmers raise livestock to make biodiesel. They raise livestock to put beef and pork and poultry on our dinner plates. Not every bit of an animal is edible, of course, and, therefore, meat processing plants produce countless tons of scrap in the form of animal fats and oils. It is waste, and it must go somewhere. The biodiesel industry puts it to good use.

The biodiesel industry creates value where it did not previously exist.

It is rather obvious that nobody is going hungry because biodiesel plants are making fuel from leftover animal fats and greases. Neither is anyone going hungry because biodiesel plants are making fuel from soybeans—but to many people that is less obvious. Soybeans certainly are valuable, as is livestock. But not everything that is inside a soybean is of equal value for food production, just as not everything inside a pig or cow or chicken is of equal value. The biodiesel industry takes what was unwanted and gives it purpose.

A Farmer's Perspective

Greg Anderson knows soybeans. A fifth-generation farmer, he works a Nebraska spread that has been in his family for 144 years. His great-great-grandfather homesteaded the original plot in 1873, and Anderson now farms five hundred acres. "It's in my blood," he says. "It's everything that I've always wanted to do."

In his state and across the country, he says, farmers are careful stewards of the land. Farmers "want that land to be productive not just in their lifetime but for many generations to come." Technology, he says, has transformed agriculture. "It's just out of this world. It's technology that has come to the farmer, a lot of it through NASA and through the military, and we are now using that to grow more food in a better way." Farmers are getting superior yields at a lower cost with much less labor and without the environmental issues of years past.

After spending hours on the tractor, Anderson finds the time to be active in agricultural and industry circles. He is treasurer of the NBB, and is a member of the Nebraska Soybean Board and the National Oilheat Research Alliance board. He is a former chairman of the United Soybean Board. He helped to establish the US Soybean Export Council, and he chaired QUALISOY, an industry-wide ini-

Greg Anderson

tiative to develop soybean traits. He has traveled to Mexico, Brazil, Argentina, and Chile on behalf of the soybean industry, and to Germany, Belgium, and other European nations to explore their biodiesel markets.

The biodiesel industry has served soybean farmers well, he says. He cited a 2016 study by Informa Economics that the industry had added sixty-three cents a bushel to the price of soybeans. That translates to about thirty-eight dollars per acre more for Nebraska—and the state has more than 5 million acres of soybeans.[6] In other words,

---

6 The National Agricultural Statisitcs Service, "2016 State Agriculture Overview: Nebraska," United States Department of Agriculture (2016): https://www.nass. usda.gov/Quick_Stats/Ag_Overview/stateOverview.php?state=NEBRASKA.

the biodiesel industry is benefiting Nebraska farmers by about $200 million a year. "If a farmer has a thousand acres of soybeans," says Anderson, "that's an extra $38,000 in his pocket." Nationwide, farmers planted 89.5 million acres in soybeans in 2017, which is up about 6 million acres from the previous year, according to the USDA, which forecast a harvest of 4.4 billion bushels.[7]

The Glut of Soy Oil

Anderson's grandfather introduced soybeans to his Nebraska farm in the 1950s, primarily for fodder. At the time, that was the value that many farmers saw in the crop before discovering the profit in letting the beans mature to sell for their protein value.

Soybeans are a legume that is related to peas, clover, and alfalfa. They are rich in both protein and vegetable oil. The crop is grown throughout the United States, but the Midwest climate and soils are ideal. A soybean plant can produce dozens of pods, each with two to four beans, which are harvested in the fall. About half of the US soybean crop is exported as whole beans to other countries. The remainder goes to processors to be cleaned, dehulled, and crushed, then it is typically is rolled into flakes.

The crushing and milling separate the oil from the meal. Rich in protein, the meal makes an excellent feed for livestock. Some of the oil is sold for food uses, such as salad dressing and cooking oil for

---

7   Teresa White and Lance Honig, "U.S. Farmers Expect to Plant Record-High Soybean Acreage," United States Department of Agriculture, National Agricultural Statistics Service (March 31, 2017): https://www.nass.usda.gov/Newsroom/2017/03_31_2017.php.

deep fried products. Most of the oil, though, would have no use were it not for the biodiesel industry.

Soybeans are not native to the United States; the beans first arrived as ballast on ships returning here from China in the early 1800s.[8] "And now it's come full circle," says Anderson. "The beans came over from China, and now we're sending them back." Soybeans are the top US export crop—"about every other row is exported"—and China is the biggest customer, using the protein-rich meal primarily to feed its thriving pork and poultry industries.

The crop is increasingly in demand worldwide as growing populations need that protein, Anderson points out, and it has come to play an important role for the US economy. "Soybeans are not just for tofu and soy milk smoothies," he says. In fact, relatively little of the crop goes directly to human consumption. Its value for food production is primarily to feed the animals that give us meat.

It was that demand for the protein that drove the growth of the soybean industry in America. Meanwhile, the surplus oil was hurting the market. "The excess soybean oil was dragging the market down," Anderson says. "Like a cloudy, rainy day, the sun just wouldn't shine. There wasn't much demand for it, but we were expanding soy production and, thus, had more and more oil stockpiling."

The farmers and the crushers didn't have the option of producing only the protein meal, and so the oil supply outstripped the market for it. When you oversupply the market, the price for the commodity falls. If the price for the oil falls below the cost of producing it, then the consumers of the protein must carry the burden and pay more.

---

8   Illinois Soybean Association, "The History of the Soybean": http://www.podto-plate.org/_data/files/Classroom%20PDFs/History_of_Soybean.pdf.

"The companies that were using the vegetable oil were paying this artificially low price for the oil," Scott says. "If the consumers of the oil are not paying for the production, that meant the consumers of the protein had to pay for the production. That was making soy protein more expensive on the market relative to other forms of protein for animal feed."

The soybean industry needed to find a way to stimulate the demand for the excess oil because it could not stem the supply of it—not unless it stopped producing the protein, too. If the farmers wanted the meal, they had to deal with the oil. One outlet that the industry found was a market for partially hydrogenated soybean oil to replace saturated fats, such as palm oil. That helped the industry for years.

"At that time, people thought saturated fats were really bad for heart health," Scott explains, so consumers viewed soybean oil as the smart choice. Researchers have since learned that adding hydrogen to soybean oil produces trans fats—"which are even worse than poly-unsaturated fats. So that was a temporary fix." Another more recent controversy arose over genetic modifications intended to enhance the quality of the soy oil. Many people believe GMO tampering has done just the opposite and rendered soy oil unfit for human consumption."

Along came biodiesel. "Biodiesel was able to take off and grow," Scott says, "because all of that excess soybean oil—which was being kicked out of the market again because of trans fats—could now be turned into fuel. Biodiesel is kind of like a safety valve that lets you bleed off that excess supply of vegetable oil into the market."

In effect, the development of the biodiesel industry set a floor for the price of vegetable oil at its basic energy value. "There's been a lot of confusion," Scott says. "People thought, *You're increasing food prices*—but the industry wasn't pushing the price of vegetable oil above its market value or above its production cost. Instead, he says, the industry was lowering the costs for protein consumers and letting the market function as it should.

In global trade, according to Anderson, soybean oil today is treated as an energy commodity by the Chicago Board of Trade and other traders worldwide. "As you track soy-oil prices over, say, the last ten years, they actually have followed the price of crude, which never happened before." The influence of the biodiesel industry is evident.

And it was the farmer, Scott emphasizes, who was the midwife of the biodiesel industry in America. It developed, in large part, as a product of agricultural economics. "In the early 1990s, I don't think there was a whole lot of thought about reducing greenhouse gases or tailpipe emissions. To my knowledge, all of those things were, in fact, coincidental. We've been miraculously fortunate that biodiesel has all of these multiple benefits. We've learned a lot in the last ten years about the sustainability of biodiesel."

Affordable Protein

The biodiesel industry, in examining its own sustainability, has asked itself some tough questions. Could the industry indeed grow so large that it could use too much of the vegetable oil and compete with food supplies?

Anderson often has heard that argument that the biodiesel industry increases food prices. "We've seen that ebb and flow with folks," he says. "The Grocery Manufacturers Association led a big charge on food vs. fuel, especially when crude oil was so high." The raw materials that should be going to make food, the critics have said, should not be diverted into making fuel.

Not much of the price of food, however, goes to the farmer, says Anderson, citing President John F. Kennedy's observation that "farmers buy retail, sell wholesale, and pay the freight both ways." Of the price of a loaf of bread, for example, a wheat farmer gets only a few cents, Anderson says. Much of the cost of food results from the manufacturing and the transportation.

Even so, US households spend only 6 to 7 percent of their income on food, less than anywhere else in the world.[9] In much of Europe, the percentage is two or three times that much.[10] In China and Russia, it's 25 to 30 percent. In Nigeria, it's more than half.[11]

Far from increasing the cost of food, the biodiesel industry lowers it, Anderson emphasizes. Here's how it works: Now that soybean farmers have a market for the excess soy oil, they have an incentive to produce more of their crop, which puts more of the protein meal on the market as well. Livestock producers prefer that source of protein, so now it can be priced competitively. They can feed their cattle, poultry, and hogs less expensively, and that translates into lower meat prices for the consumer. It's Economics 101.

9   Annemarie Kuhns and David Levin, "Food Price Outlook," *USDA Economic Research Service* (July 3, 2017): https://www.ers.usda.gov/data-products/food-price-outlook/charts/#expenditure.

10   Alex Gray, "Which countries spend the most on food?" *World Economc Forum* (2015): https://www.weforum.org/agenda/2016/12/this-map-shows-how-much-each-country-spends-on-food/.

11   Matthew Boesler, "Here Are the Countries that Spend the Most on Food," *Business Insider* (February 24, 2014): http://www.businessinsider.com/countries-that-spend-the-most-on-food-2014-2.

Nonetheless, critics have declared that biodiesel production is increasingly claiming the calories that the world's growing population will need. Scott points out, however, that the calories from carbohydrates and fat are not enough. Protein is essential to our diet, and nutritionists say it should represent 30 percent of the calories that we consume. Our bodies cannot function without protein, and the growing populations will need ever greater amounts of it. Protein also is the most expensive part of our diet, and, therefore, most people ingest far too many of their calories in the form of carbohydrates and fat. The key to an affordable and healthful food supply is to lower the cost of the protein.

Originating from plant life is the fundamental nature of food— including meat, a top source of protein. All plants contain more fats and carbohydrates than protein. To collect enough protein to properly feed the world, we therefore will also harvest far more carbs and fat than we can use in our food supply. A soybean, for example, contains 34 percent protein, and 66 percent fat and carbs.[12]

What, then, should we do with that excess? Throw it away, or find another use for it? Those carbs and fats are energy from the sun, Scott says, which plants are masters at storing. In fact, when we burn fossil fuels, we are releasing the solar energy that was stored in plants eons ago. The biodiesel industry gives us the opportunity to produce a fuel from the solar energy stored in the plants that we grow today rather than continuing to depend entirely on fossil fuels as the source of that energy.

---

12   Don Scott, "Biodiesel Complements the Food Supply," The National Biodiesel Board (February 28, 2017): http://www.biodieselsustainability.com/2017/02/28/biodiesel-complements-food-supply/.

The biodiesel industry has been rising to the challenge. Every year, says Anderson, it has been using about 6.2 billion pounds of soybean oil, which is about 30 percent of all the oil coming from the domestic crush. "Those are numbers we could only dream about ten years ago," he says. Although the industry does use other feedstocks, at least half of the biodiesel production in America originates with the soybean.

> *The biodiesel industry gives us the opportunity to produce a fuel from the solar energy stored in the plants that we grow today rather than continuing to depend entirely on fossil fuels as the source of that energy.*

From whatever source, we can make biodiesel in full confidence that we are not diverting food or robbing calories from a hungry world. Instead, by giving a value to those excess fats and carbs, the biodiesel industry makes the protein less expensive. The world can afford to eat better.

Using Our Land Wisely

Will we use up all our land in the process? Some conservationists see biodiesel production as an incentive for farmers to till up more and more acreage to feed the refineries. Too many wildlife habitats have already been sacrificed to agriculture, they say. On the contrary, Scott says, "the crops we grow and the way we grow them is naturally minimizing the footprint of agriculture. Crop yields have increased astronomically because farmers are getting good at what they do. We have the best farming technology in the world."

Soybeans produce an extraordinary amount of protein per acre. All that protein goes into the food supply. To get that much from any other crop would require planting many additional acres. Soybean farmers are growing an efficient crop that requires less land to meet the demand for protein—and with advances in technology, they are boosting the yield even further. Much of that increased yield comes from enhancing growth through a better understanding of nature. For example, farmers have learned that they can increase efficiency by double cropping—growing wheat in the winter and soybeans in the summer, for example, optimizing the productivity of the same acreage.

A unique aspect of soybeans as a legume crop is their ability to "fix" their own nitrogen. Nitrogen is a major component of amino acids, the building blocks of proteins. Corn requires nitrogen fertilizer to replace what it pulls out of the land, and the runoff from that fertilizer often is criticized as harmful to the environment. Soybeans, however, pull nitrogen right out of the atmosphere. A microbe on the roots extracts the nitrogen and converts it for use by the plant. In effect, soybeans make their own fertilizer. The farmers do not have to add synthetic fertilizer. They will typically grow corn and soy in rotation. The soy crop will fix nitrogen into the soil and use it for its own growth. The next year, some of that nitrogen will carry over to nourish the corn crop. The soy crop reduces the amount of fertilizer that the farmer must use.

More protein, less fertilizer, less land to cultivate—that's a winning combination. Those advantages make soybeans a crop worthy of promoting, and the biodiesel industry does just that. When farmers know that they will have a market for the soybean oil, they feel encouraged to plant the crop for the protein. "We don't grow soybeans just for biodiesel," Scott emphasizes. "We grow soybeans for protein, and we use the leftover oil for biodiesel."

By encouraging soybean farming, he adds, we get a protein-rich feed for animals—meaning we can raise more livestock on less pasture land to satisfy humanity's need for a balanced and nourishing diet. "Biodiesel is part of all this. We're helping incentivize efficient protein production with soy."

The world's growing population will be needing that protein. People will be eating more protein than ever before as standards of living increase globally. It is our responsibility as world citizens to use the resulting surplus of fats and oils to our best advantage and not squander their potential. Far from subtracting from the food supply, biodiesel production adds to it. For every gallon of biodiesel produced from soybean oil, the industry contributes thirty pounds of protein and twenty-two pounds of carbohydrates and dietary fiber.[13]

> **_Biodiesel is a friend of agriculture, a friend of land and wildlife conservationists, a friend of environmentalists, and a friend of consumers._**

Soybean production lets us feed more people while farming less land, saving more of it for natural habitats. The biodiesel industry clearly has proved itself to be a good partner. It is a friend of agriculture, a friend of land and

---

13   Don Scott, "Biodiesel Complements the Food Supply," _Biodiesel.org Sustainability Blog_ (February 28, 2017): http://www.biodieselsustainability.com/2017/02/28/biodiesel-complements-food-supply/.

wildlife conservationists, a friend of environmentalists, and a friend of consumers.

Solution for Surpluses

American farmers produce far more than the nation can consume, says John Campbell of Ocean Park Advisors, the biodiesel pioneer who built the industry's first commercial plant. "We have 5 percent of the world's population, and we have 10 percent of the world's agricultural production, in rough numbers," he says. "So, we are always looking for ways to export and get rid of the surplus."

In the past, agricultural policy was designed to pay farmers not to produce, and the government tried to give away the surplus in food-aid programs. "We learned in the 1980s that it was a policy that was just not sustainable," Campbell says. "We couldn't pay farmers enough to not produce. We just kept losing market share" to other nations, which filled the void. "We were just getting farther and farther behind. In 1983, farmers were paid to idle 80 million acres of land. That's how bad it was. We had so much surplus of everything—and there was just no place to put it."

The rise of the biofuel industries represented a change in direction toward looking for productive new markets for the surplus. "If we don't do ethanol and biodiesel and those things, what are we going to do with all this stuff? We can't eat any more of it; we're full. The productive capacity of our farmers far exceeds the eating capacity of Americans. Rather than just relying on exports, why not build a domestic value-added industry? Why not turn these surpluses into fuel oil?"

The biodiesel industry has done precisely that, and it serves as one more example of the power of human ingenuity. Businesses thrive by providing products or services that people want and that fill a need in our society. The ultimate efficiency is to find a new use for what previously had been unwanted. As we have seen, the petroleum industry did that with both gasoline and diesel fuel. The biodiesel industry is doing that with the waste from meat-processing plants and with the excess oil from soybean processing.

And, in turn, other industries benefit from the byproduct of biodiesel refining: all that glycerin does not go to waste; it finds a market. It is a valuable ingredient of pharmaceuticals, antifreeze, airplane de-icing fluid, and more, and researchers continue to find other applications. They are also looking for other uses for biodiesel besides fuel, such as renewable lubricants and cleaning agents. Ingenuity is alive and well in the world of business.

Since soybeans are grown primarily for the protein, and consumers cannot use all that vegetable oil, where will the excess go? The biodiesel industry provides the answer: convert that unwanted product into energy to power trucks, tractors, cars, buses, trains, and boats. This is energy from waste—and we must not waste this opportunity to efficiently, safely, and sustainably meet our nation's needs.

Joe Jobe, the former NBB chief, puts it this way: First, you crush a soybean, and out comes protein and oil. You feed the protein to a chicken, and make biodiesel from some of the leftover oil. The chicken gets plump, with plenty of meat and fat. You turn that fat into biodiesel, and then you fry the chicken with some more of the leftover oil. Then you turn the used cooking oil into biodiesel. You eat the chicken, and start all over again.

"You start out with some biodiesel to grow that first batch of soybeans," he says, "and you end up with five times the biodiesel from three different sources, and an array of renewable co-products—and a delicious chicken dinner. Who else does that? How awesome are we?"

CHAPTER 7

LEVELING THE PLAYING FIELD

Where gauchos on horseback once herded cattle on the Argentine pampas, today endless acres of soybeans are the lay of the land. And, over the past decade, biodiesel plants have been sprouting along the rivers, too. Argentina, like the United States, has its eye on biodiesel and its prosperous prospects.

Unfortunately, the Argentine biodiesel industry has been based on the potential for export, and much of that nation's biodiesel has shipped out to American ports. Not only does Argentina subsidize its own production through tax incentives, but US tax policy in recent years also encouraged such offshore production. In effect, US policy has supported Argentine jobs instead of American ones. If the objective was to reduce imported petroleum, what was the sense in replacing it with imported biodiesel?

As the imports swallowed nearly half of the domestic demand in 2016, American producers felt the pressure. Most of the imports were originating in Argentina, with much of the rest coming from Indonesia and a small amount from Finland, Germany, in addition to reciprocal trade with Canada. "This has obviously challenged our domestic producers," says Donnell Rehagen, the NBB chief executive. "They were forced to compete with an imported product that is subsidized and dumped. The pricing competition has been extremely difficult."

According to government data, the United States imported 708 million gallons of biodiesel in 2016. Two-thirds of that came from Argentina, with much of the rest from Indonesia. The NBB reports that by the end of 2016, imports from Argentina (which makes biodiesel from soy) and from Indonesia (which makes it from palm oil) had increased 464 percent in just two years. That happened after the EPA in 2015 allowed the imports to qualify for US credits. The NBB viewed this as illegal dumping—at 23 percent below market value for the Argentine imports, and at 34 percent below for the Indonesian biodiesel—and pushed for the imposition of federal trade duties to stem the tide.[14]

"We were dealing specifically with Argentina and Indonesia," Rehagen points out, "but that could just as easily be China and India if they ever decide to get into significant renewable fuel production. The players could change down the road, but the concern will always be the same."

The Argentine double dipping on production incentives is a prime example of why the United States needs to pursue policies that will

---

14   Rosemarie Calabro Tully, "NBB Fair Trade Coalition Succeeds in Latest Stage of Biodiesel Import Case," *National Biodiesel Board* (August 22, 2017): http://nbb.org/news/nbb-press-releases/2017/08/22/nbb-fair-trade-coalition-succeeds-in-latest-stage-of-biodiesel-import-case.

level the playing field for the domestic biodiesel industry. Federal strategies should be designed to allow US producers to compete in a fair marketplace so that our homegrown biodiesel can fulfill its exciting potential. If it is in our nation's interest to promote environmental integrity, economic growth, and energy security, then certainly it is in our best interest to advance policies that will help the biodiesel industry instead of hindering it.

> *If it is in our nation's interest to promote environmental integrity, economic growth, and energy security, then certainly it is in our best interest to advance policies that will help the biodiesel industry instead of hindering it.*

"A Match Made in Heaven"

US biodiesel policy has its roots in the Energy Policy Act of 1992, enacted before the industry existed. "That was one of the reasons why the biodiesel industry started," says Steve Howell, the NBB senior technical director. "It basically said that government fleets have to use some sort of alternative fuel because the government was a big purchaser and was going to lead the way." In addition, private companies that sold an alternative fuel, such as electric companies and natural-gas utilities, also had to purchase alternative fuel vehicles. Later, when biodiesel became readily available, a modification of the act allowed those fleets to comply with it by using a 20 percent blend in their existing diesel engines. The fleet operators found that to be a far more economical way to meet the mandates of the Energy Policy Act.

"That was a major turning point," Howell says. "That's when commercial fuel sales really started happening, when government fleets could use B20 and meet their requirements. That's still in place today, and government fleets still use a lot of biodiesel as their mechanism. They buy normal diesel vehicles and use B20 in them, and that counts the same as buying an alternative-fuel vehicle."

Other breakthroughs for the industry, he says, included EPA registration of biodiesel based on extensive health-effects studies, and the adoption of the ASTM technical standards needed so that engine manufacturers could approve the blends for use in their equipment. In other words, the industry got biodiesel designated as a legal fuel, and it made strides toward building market confidence. Much of that was happening around the turn of the millennium.

In the years since, two federal programs have had a substantial influence on the advancement of the American biodiesel industry. One is a tax incentive for those who blend biodiesel into the fuel supply, and the other is the Renewable Fuel Standard (RFS). Those measures, along with a variety of policies, incentives, and mandates at the state level have given the industry the footing it needs to climb a difficult slope. The intent of the government programs has been to encourage production of biodiesel, promote competitive pricing at the pump, and offer it as a cost-effective means of meeting environmental regulations.

"Our industry would not be the size it is today, with the capacity that it has, were it not for the biodiesel tax credit," Rehagen says. "The tax credit dramatically changed how biodiesel was produced, and the scale at which it was produced. Then the RFS legislation created a very steady demand for biodiesel. The timing of all that coming together—a tax credit that incentivized larger-scale production, and a standard that in some ways mandated the use of renewable fuels across the infrastructure—was kind of a match made in heaven."

A Global Perspective

The biodiesel industry is essentially an oil industry, says Doug Whitehead, the NBB's chief operating officer. "We are a liquid-energy source, made in America, from American ingenuity, and so we are an oil industry. We've added refining capacity all over our country—new

capacity—and if we can continue to get positive signals from the state and federal governments, then those signals will be heard loud and clear, and our guys will continue to invest and add jobs and keep working toward reducing our dependence on foreign oil."

Doug Whitehead

As a liquid fuel, biodiesel competes on the world energy market along with petroleum-based fuels. Since 1973, crude oil prices have spiked and plummeted in a range from less than $20 to over $140 a barrel.[15, 16] "You can imagine what that does to industries that rely on that energy," Rehagen says. "We get pulled in different directions in the same manner." The Organization of Petroleum Exporting Countries

---

15   The Energy Information Administration, "Crude Oil Price History," *FedPrimeRate. com* (October 13, 2017): http://www.fedprimerate.com/crude-oil-price-history.htm.

16   "Crude Oil Prices - 70 Year Historical Chart," *Macrotrends* (Accessed October 17, 2017): http://www.macrotrends.net/1369/crude-oil-price-history-chart.

(OPEC) has the power to raise or reduce prices on a whim, "and we just have to take it."

Price fixing is a fact of life in global-oil economics. "It's illegal in the United States, but that's the market that we compete in. It's very challenging." For example, when OPEC saw that hydraulic fracking had become a viable option in the United States, "what did they do? They significantly dropped the price of oil and forced a lot of those industries to be sidelined. That is the reality of the environment that we operate in, 'we' being biodiesel as well. They saw that all these jobs were being created in the domestic oil industry, and we were reaching some higher levels of energy independence, and OPEC decided that's not going to happen."

In his many years at the helm of the NBB, Joe Jobe heard it all, including this refrain: *Biodiesel should stand on its own in the free market, unfettered by energy policy.* "Sounds great," he says, "but the truth is, nothing is unfettered about the energy markets." He, too, has pointed to the manipulation of petroleum production and supply by OPEC. Its goal: keeping prices artificially high over the long haul, with intermittent price drops to thwart emerging alternatives. Even though the World Trade Organization outlaws market fixing, he says, member nations have been reluctant to prosecute members of the cartel as a matter of foreign policy.

As a result of that price uncertainty, says Rehagen, an investment in any kind of energy production, including biodiesel, is risky. "As you can imagine, the economic dynamics of building a biodiesel plant in Iowa or Illinois or Kansas or New York or California is extremely challenging. The biodiesel tax credit and the RFS have created a mechanism where there's a little bit more certainty to that." They paved the way to increased production and use of renewable fuels—"exactly what Congress intended."

Many roads are paved with good intentions. The policymakers may lose sight of the objective, or they may be reluctant to make the necessary adjustments, or a new crop of policymakers may see things differently. Sometimes they never fully grasp the significance of the breakthrough that they helped to accomplish. Rehagen is quite familiar with the political arguments. "We have members of Congress who ask us, 'So when are you not going to need this tax credit?' They don't understand the global dynamics. To me, the answer would be: 'I don't know. When will Congress have something to say about oil prices? That's probably about the time.' And we know that's never going to happen."

A growing economy needs more energy. It is in the best interest of every citizen that the government engage in incentives for domestic energy production, including alternative energies—biofuels, wind, solar, and whatever other new opportunities arise. The global forces may be beyond our control, so it's what happens at home that makes all the difference. Energy is essential to our prosperity. The United States must seize the moment or risk losing it.

> *The global forces may be beyond our control, so it's what happens at home that makes all the difference. Energy is essential to our prosperity. The United States must seize the moment or risk losing it.*

Whitehead extols his organization's role in meeting the challenges. "Two of the biggest lobbies in the world are agriculture and oil, and often that was kind of a bumpy relationship because they're both seeking federal policies and tax programs and those sorts of things. I'm very proud that the National Biodiesel Board, its leaders, staff, and contractors have remained a single, coordinated voice focusing on how much more good we can do together."

To survive, the biodiesel industry must be ever vigilant that the policies designed to help it compete and grow remain effective. As Whitehead puts it, "Administrations change, members of Congress change, leaders change, the economy changes. There are so many factors." And the industry must continue to demonstrate that it delivers on its promises.

The NBB has steadfastly advanced the conversation on the benefits of biodiesel and the wisdom of the tax credit and the RFS. Let's take a closer look now at those two crucial federal programs that have been driving the US biodiesel marketplace.

The Biodiesel Blenders' Tax Credit

Passed by Congress and signed into law by President George W. Bush, the biodiesel tax credit took effect in January 2005. "That was a very interesting time for the biodiesel industry," Rehagen says. "At that time, there were maybe ten companies in the whole country producing biodiesel. There was a lot of potential, but there were lots of barriers as well. It did exactly what it was intended to do, which totally changed the economics of building a biodiesel plant in the United States."

By that August, he says, numerous companies were already getting into the biodiesel space, developing business plans all around the country, and today almost every state has biodiesel plants. "We have to give a lot of credit to the biodiesel tax credit for changing those dynamics back in the mid-2000s," Rehagen says. The tax credit led to the building of bigger production facilities—among them, the HERO BX plant in Erie, with an annual production capacity at 50 million gallons. Previously, a plant with a capacity of 15 million gallons would have been considered large. After the tax credit, many of the newer plants had a capacity of twice or three times that much, "and today we have plants that are over 100 million gallons."

President George W. Bush signs the bill in 2005

Congress established the tax benefit as a $1 per gallon credit for facilities that were blending biodiesel into the fuel supply, regardless of who made the biodiesel. Over the years, Congress at various times has allowed the credit to lapse before reinstituting it. It lapsed at the end of 2009, 2011, 2013 and 2014, generating uncertainty throughout the industry into the ensuing years. Each time, Congress eventually reinstated it retroactively, but each delay also gave pause to the biodiesel industry. It lapsed again in 2016 as the industry, ever hopeful of a longer term legislative solution, awaited action from the new Trump administration.

The uncertainty shakes the market and stifles investment. "How can you price your product," explains Rehagen, "when you don't know if that tax credit is going to be available or not be available?" Biodiesel producers, he adds, have held back on investing in their facilities to enhance production because they are unsure whether the federal government will be supporting them long term.

The tax credit has been an attractive incentive for the blenders, of course, and encourages their continued output, but it has treated imports on a par with US-produced biodiesel—which explains why offshore sources captured so much of the market. The US industry, with the backing of the NBB, has been advocating that the blenders' credit be changed to a producers' credit to further stimulate invest-ment in American biodiesel plants. They want the words "domestically produced" to truly mean *made in the USA*, not just *mixed in the USA*.

"The goal of the tax credit was to grow the US biodiesel industry, not the international biodiesel industry," Rehagen says. "We've been asking Congress to see the tax credit for what it is, an incentive for biodiesel imports. There's plenty of idle capacity in the plants that have been built in the United States to fill the demand that's currently being met by imported product."

Argentina's system of differential export taxes encouraged the conversion of soy oil into biodiesel to be shipped abroad. In the past decade, the gauchos have been fading fast into folklore. The cattle ranches have been yielding to the soybean fields. The rural lifestyle is changing as industrial agriculture has swept away many of the smaller- and medium-sized farms of the grasslands.[17]

Only a small percentage of the nation's soy crop is consumed there, however. Almost all of the production is destined for overseas markets. Argentine investors also soon saw the profitable potential for an overseas market for biodiesel. Argentina's favorable export-tax treatment for the fuel led to a bumper crop of biodiesel refineries that soon began meeting the US demand.

> *We're fine with competing globally, but when you have a country such as Argentina that's incentivizing the biodiesel produced there, and then the US taxpayers are incentivizing it again, that creates a very unlevel playing field for our domestic producers here. Most energy products are traded on the world market—but it should be a fair market.*

And here in the USA, the blenders' credit helped to fill the pockets of the Argentine producers. "That was never the intent of it," Whitehead says. "Our intent was a domestic program, made in America, used in America. The unintended consequence was that smart accountants found the loophole that companies from outside the United

---

17 Nicholas Kusnetz, "Change on the Pampas: Industrialized Farming Comes to Argentina," NACLA.org (2014): https://nacla.org/news/change-pampas-industrialized-farming-comes-argentina.

States could import their fuel and take advantage of that blenders' tax credit. We certainly didn't want our producers here to get beat by imported fuels. That's unacceptable. We are for fair trade, and by definition, fair is fair. Our guys love to compete, all day, every day—and historically they win, but it's hard to win if you're ten cents to thirty-five cents at a disadvantage because of double subsidies."

It's not as if the United States is unable to meet its own needs. "We're very confident that if the demand is there for higher volume of domestically produced biodiesel, we can definitely meet that," Rehagen says. "If there's going to be an investment of taxpayer dollars in an industry like biodiesel, it should create reward for the US businesses. That's what we're looking for. We would love to put to work some of the existing capacity that's not being utilized. We know that's going to result in more jobs at a lot of those plants." If the industry is to grow and add those jobs, he says, the producers need a sense of certainty that the demand will be there in the marketplace—and that's what the tax credit and the RFS should be consistently promoting.

When the tax benefit was implemented as a blenders' credit in 2005, Rehagen says, it functioned as an offset to the taxes that fuel dealers already owed on petroleum diesel. The policymakers at the time felt it made sense to apply the credit at the level where the tax was currently being levied, giving dealers a break if they chose to blend biodiesel into the supply. A producers' credit would apply the benefit at the point of manufacture instead of at the point of blending. The NBB and other advocates of the change point to the significant increase in biodiesel imports in recent years and how the Argentine producers took advantage of US policy treating them as equals.

"Imports have absolutely surged," Jobe says, "and soon half of the volume is going to be imports—in fact, we are already pretty much there. That's a troublesome policy. You have an incentive designed to stimulate homegrown biofuel production, and half of it is going to subsidize foreign producers. To make it a producers' credit would very effectively deal with that."

The Renewable Fuel Standard

With the aim of reducing greenhouse gas emissions, expanding the renewable fuels sector, and reducing reliance on imported oil, Congress authorized the Renewable Fuel Standard (RFS) under the Energy Policy Act of 2005. Originally, the focus was on ethanol. Two years later, the Energy Independence and Security Act expanded the program, and this time it included biodiesel.

Each year the Environmental Protection Agency (EPA), which administers the RFS, holds public hearings and establishes the minimum amount of renewable fuels that must be blended into the petroleum supply. It sets separate levels for ethanol and biodiesel. Biodiesel is categorized as an "advanced biofuel" that reduces greenhouse gas emissions by at least 50 percent. For 2017, the EPA set nationwide levels of 2.1 billion gallons of biodiesel and 4.28 billion gallons of advanced biofuels.

Though the EPA rules on those volume levels, all the branches of government influence the process. On behalf of their constituents, the legislators every year push for either higher or lower volumes, or various regulatory changes. And every year there have been judicial challenges as stakeholders on both sides line up in court.

The program designates "obligated parties"—namely, the petroleum refiners and importers—who must incorporate their share of the renewable fuels into the supply. They can either purchase the biofuel directly and blend it themselves, or they can buy credits called renewable identification numbers (RINs). The RIN system basically is a means of tracking who is using how much of the fuel. When a refiner buys a RIN, in effect it is paying somebody else to do the blending. The system is intended to provide flexibility. If a refiner, for example, is far away from a biodiesel plant, it still can meet its obligation by purchasing credits to fulfill the mandate.

The combination of the RFS and the federal tax credit helped the biodiesel industry tremendously. By 2010, the standard was getting into full gear and the industry was seeing the benefits of the governmental support. Not only did it have a tax incentive, but now it also had a federal mandate that the country would use a specified amount of biodiesel. The NBB continues to focus on preserving and improving those two policies.

Others focus elsewhere. Though some segments of the petroleum industry are supportive of the RFS, some stand firmly against it. "The anti-RFS folks," Jobe says, "are represented primarily by the large, integrated petroleum companies represented by API, the American Petroleum Institute. And they continue to oppose the RFS at all three branches of government, and to the public at large, but they have largely been unsuccessful."

The NBB has been at the forefront in addressing what most of our national leaders have for decades recognized as a threat to the entire US economy: the dangerous addiction to a single source of transport fuel. Finally, with the RFS, the federal government laid out a bold plan to deal with that addiction, and the program has been succeeding—but not without conflict between those who want to make it work and those who want to make it fail. It has been a struggle, Jobe says, in which the facts have sought to prevail against alarmist rhetoric.

Jobe emphasizes that he is not making a sweeping attack on the petroleum industry, where a lot of talented people are doing great things. "Many segments of the petroleum industry have supported the goal of diversifying our energy portfolio. The heating-oil industry is a good example, but also petroleum distributors, blenders, pipeline companies, and others have embraced biodiesel, the RFS law, and the objective of diversification." His criticism, he says, is only against those

Joe Jobe

who distort the truth in their opposition to the RFS. Specifically, he charges that RFS opponents have shown a pattern of misrepresenting the facts. They have tried to minimize the biodiesel industry's production capability, for example, despite data clearly establishing that the industry met or exceeded its volume goals.

Several years ago, the API released a study predicting that the RFS would lead to what it called the "diesel death spiral." It is another example, says Jobe, of unconscionable pseudoscience. The report spread the myth that if the biodiesel blends exceeded a "wall" of 5 percent, the RFS within a few years would drive diesel prices catastrophically high and collapse the economy. Diesel soon could cost fifteen dollars a gallon, and then seventy dollars. The API even testified about the imminent collapse before Congress. None of that, of course, happened—and yet, Jobe says, the API did not change its stance or disavow the study. In fact, it updated it: diesel could soon cost $103 a gallon, the study warned. Instead, prices fell, even as biodiesel production steadily increased.

Those who disseminate such myths, Jobe says, "have zero credibility on the RFS. They are pandering to fear and anger and peddling confusion and disinformation, and it is shameful." The success of our democracy, he says, depends on truthfulness in civil discourse. Every year, the biodiesel industry must arm itself with solid research and the facts to fight for reasonable, sustainable growth in the RFS, Jobe emphasizes. And every year, the industry must be prepared to face the outrageous and desperate tactics of those seeking to repeal it.

The RFS is a robust program that can withstand the political winds. It is embedded in permanent law via the Clean Air Act, which is a particularly difficult statute to modify, Jobe says. Neither political party has wanted to risk opening the door to changes that might be averse to their interests. The law has stringent compliance requirements and penalties. "So, because of that," Jobe says, "the obligated parties have always complied with the RFS to the best of their abilities, and they will continue to do so. These are publicly traded companies and they don't screw around with violating the law. They may oppose the law, but they comply with it."

As permanent law, the RFS does not expire like the tax credit, and, in years to come, the focus of growth will be biodiesel. As of 2017, the ethanol category reached its statutory limit of 15 billion gallons a year, and it is ethanol that has generated most of the biofuels controversy. Because the ethanol volume is now capped, all the growth will henceforth be in the advanced-biofuel category, primarily biodiesel.

"As long as we have a functioning, healthy, growing RFS, the industry will be fine," Jobe says. "The industry will survive and thrive, because the RFS really determines the number of gallons nationwide that are consumed." The biodiesel tax credit is an important boost for profit margins, he says, but the RFS mandate and the RIN credit program serve as the "shock absorber" between the price of fossil fuel and the price of biodiesel.

The Renewable Fuel Standard is not going away. It has withstood annual challenges and prevailed. The system itself is robust, but the question is whether the EPA will continue to set robust volumes for biodiesel growth in the years ahead. A lack of sufficient growth would thwart investment, and the industry would stagnate.

Much is at stake. The nation needs energy diversity. Without diversity, an investment portfolio is weak and vulnerable. Such is the case with our energy portfolio as well. Our nation needs a strong and comprehensive strategy that builds our domestic petroleum resources and diversifies the energy supply. The American biodiesel and petroleum industries must work together as partners, not adversaries, to advance that strategy. The RFS has been a good start. It works, and it can work better.

> *It doesn't have to be buttons vs. zippers ... buttons and zippers are better together. And biodiesel and diesel fuel are better together, too.*

Initiatives at the State Level

In addition to the federal programs, some states have passed their own legislation to give a boost to biodiesel. Some have mandates requiring a certain volume or a particular level of blend; some offer tax incentives of various types, and some have issued environmental standards for low-carbon fuel.

Shelby Neal

"There's a tremendous amount of diversity," says Shelby Neal, the NBB director of state governmental affairs. The state policies alone guarantee production of about a billion gallons a year, he says, even if the federal support were to evaporate. Each state has its own reasons for backing biodiesel, and each has its own priorities and approach. Texas and California, for example, have both made strides in developing policies that support biodiesel, but their history and their politics have colored how they go about it.

"California and Texas are separated by a thousand miles and a hundred years," Neal says. They're really different places. Texas is politically about as conservative as you can get, and it's about as pro-business as you can get. And then you look in California, and that's as progressive politically as you can get. What they value are social benefits and environmental benefits of products."

The three principal reasons for using biodiesel are its environment benefits, its economic benefits, and its contribution to energy security. "Those are all reasons that those two states have significant policies, but their priorities are vastly different," Neal says. "In Texas, number one is the economics. They have more biodiesel-production capacity than anywhere else. If you can make money on it, they'll do it in Texas." That's not the top priority in California. "The environment is number one there."

A state's social and political climate influences the policies that it chooses, and three primary approaches have emerged across the nation. Some states have issued mandates requiring blends at certain levels, some have offered tax incentives to stimulate investment and consumption, and some have set low-carbon fuel standards. "These are three really good options, all of which have worked well, and so really which option is chosen is a function of the politics and the history of individual states and regions," Neal says.

"In general, we'd prefer not to use the word mandate, but most people use it," Neal says. "I think a more appropriate way of explaining it is that is you're just enhancing the current fuel standard to make it cleaner and more domestic and more renewable."

By any name, the mandates work. Minnesota, which has three biodiesel plants with a combined production capacity of about 63 million gallons, was a pioneer of the mandate approach. The legislature there passed a law in 2002, several years before the federal biodiesel initiatives, requiring that diesel fuel sold in the state contain at least a 2 percent blend. The law was implemented in 2005. The state took a bolder move in 2014, increasing the requirement to a B10 blend from April through September and B5 for the colder months. The state plans to move to a B20 mandate in May 2018.

"They were the first to do a mandate," Neal says, "and, as you might imagine, a lot of people didn't like that. The fuel industry is basically a zero-sum game, so if 2 percent biodiesel goes in, that means 2 percent petroleum comes out. There was a lot of opposition to that, but not from the public—from special-interest groups. There's been a lot of work to make sure that that policy has been successful." Minnesota's success with its program paved the way for policies in other states, says Neal, such as Oregon's B5 requirement and Pennsylvania's B2 requirement.

Pennsylvania's law, signed in 2008, carried a provision that the mandate would commence as soon as annual biodiesel production in the state reached 40 million gallons.[18] The HERO BX plant in Erie soon met and exceeded that requirement, and in 2010 the mandate went into effect. It required a 2 percent blend in all diesel sold in the commonwealth. The law will require a 5 percent blend once biodiesel production in the state reaches 100 million gallons a year, a 10 percent blend at 200 million gallons, and a 20 percent blend at 400 million gallons. In addition, Pennsylvania lawmakers have been considering legislation for a B2 mandate on home heating oil sold in the state. Rep. Patrick Harkins of the Erie district introduced such a bill in the 2017 session.[19]

Neal also praised New York City's decision to progressively increase Bioheat blends to 20 percent by 2034. It was the first major jurisdiction to implement a Bioheat requirement. "New York City is bigger than most states, and they have an oil-heat market of a billion

---

18 Erin Voegele, "Pennsylvania biodiesel mandate to take effect," *Biodiesel Magazine* (January 15, 2009): http://www.biodieselmagazine.com/articles/3178/pennsylvania-biodiesel-mandate-to-take-effect.

19 Pennsylvania House Bill 1306, Legiscan (2017): https://legiscan.com/PA/text/HB1306/id/1606022.

gallons, roughly," he says. "They have implemented it, and they have not had any issues whatsoever. There have been no supply disruptions. There have been no performance issues. There have been no pricing issues." As a city of international scope, New York has been the perfect proving ground to demonstrate how well biodiesel can work. *If New York can do it,* advocates say, *so can you.* Neal says the city set a precedent that has been a big help in the effort to develop state policies. Case in point: the New York state assembly's legislation to expand upon the city's initiative by requiring at least a B5 blend of Bioheat in home heating oil in the downstate counties of Nassau, Suffolk, and Westchester.

Incentives such as tax credits and exemptions to stimulate consumption have been another successful approach that states have used to encourage the biodiesel industry. Fundamentally, the challenge is not that biodiesel is more expensive. Its pricing is typically competitive with or lower than the price for petroleum. The issue, says Neal, is market access. "We are like Pepsi trying to convince Coca-Cola to sell our product in their vending machines," he says. "If we look to the integrated oil companies, there's almost nothing that would cause them to sell our product outside of some sort of government intervention. So you have to do something."

One thing that government can do is create a significant financial incentive—"something where biodiesel pricing is just so attractive," Neal says, "that the petroleum-based companies would be crazy not to use biodiesel." He says Illinois was the leader in that approach. In 2003, the state began offering a complete sales-tax exemption on blends above B10, and a partial exception of a fifth of the sales tax for lower blends. As a result, most of the blends sold in the state at retail fuel outlets have been at B11 or higher.[20]

Other states followed Illinois in adopting various kinds of tax incentives. Here are a few examples: Texas exempted the biodiesel portion of blends from its diesel fuel tax.[21] New York offered a tax credit for producers of fifteen cents per gallon after a facility has made forty thousand gallons of biodiesel in a year.[22] Oregon offered an exemption from its state fuel excise tax for blends of B20 or higher in which the biodiesel is derived from used cooking oil.[23] Iowa offered retailers a state income tax credit of four and a half cents per gallon for blends of at least B5, and starting in 2018 a credit of five and a half cents for blends of at least B11.[24] "The incentives have to be significant," Neal says. "A penny a gallon isn't enough incentive to get somebody to change what they've been doing for 100 years."

The third major form of state initiative that has promoted biodiesel is clean-air policies—namely, the low-carbon fuel standards such as those that California and Oregon established to encourage the use and production of cleaner-burning fuels and to reduce greenhouse gases. California is requiring at least a 10 percent reduction in the carbon intensity of transportation fuels by 2020.[25] Oregon seeks

---

20   "Biofuels Tax Exemption: Illinois," US Department of Energy (2017): https://www. afdc.energy.gov/laws/5697.

21   "Diesel Fuel Blend Tax Exemption: Texas," US Department of Energy (2017): https://www.afdc.energy.gov/laws/5641.

22   "Biofuel Production Tax Credit: New York," US Department of Energy (2017): https://www.afdc.energy.gov/laws/6302.

23   "Biodiesel Tax Exemption: Oregon," US Department of Energy (2017): https:// www.afdc.energy.gov/laws/11062.

24   "Biodiesel Blend Retailer Tax Credit: Iowa," US Department of Energy (2017): https://www.afdc.energy.gov/laws/6082.

25   "Low Carbon Fuel Standard: California," US Department of Energy (2017): https:// www.afdc.energy.gov/laws/6308.

the same reduction by 2025.[26] The low-carbon fuel standards have stimulated a dramatic increase in biodiesel production. In California, Neal points out, the biomass-based diesel production was only 14 million gallons in 2010. Seven years later, it was 410 million gallons, "so it has been huge growth with those policies."

A lobbyist by profession, Neal says his role with the NBB requires very little lobbying. "I'm in the rare position of having a product that basically sells itself. Who doesn't want cleaner air? Who doesn't want more jobs? Who doesn't want more reliance on domestic sources of fuel?" Neal worked several years on Capitol Hill before moving back to the Midwest, where he became a lobbyist for the Missouri Soybean Association, and then he worked as a policy analyst for the governor's office. "Eventually I found my way to biodiesel," he says, "but the issues that I've handled throughout were agriculture, economic development, and environmental issues, and so biodiesel is really the confluence of all those."

It's his nature, he says, to root for the underdog. He began working for the NBB in 2008 and has seen the biodiesel industry blossom. "To see it grow from three hundred million gallons to almost three billion gallons today has been really satisfying." The dominant petroleum industry hasn't been able to overwhelm the underdog, he says, and "it's because biodiesel is just the right thing to do."

Fairness in Subsidies

Biodiesel advocates often point to the federal subsidies that Congress has given to the prosperous and powerful petroleum companies

---

26   Clean Fuels 101, www.oregon.gov: http://www.oregon.gov/deq/aq/programs/
     Pages/Clean-Fuels-101.aspx.

over the last century as a prime example of political favoritism. The oil and gas sector receives billions of dollars a year in tax breaks, some in deductions for depreciation that can be hard to track.[27] The companies can even claim "intangible drilling costs" as tax write-offs.

"The oil industry has had decades of subsidies buried in the tax code, some obvious, some not so obvious," Whitehead says. Unlike the petroleum interests, the biodiesel industry is a relative newcomer striving to establish itself. "We're trying to create this American product to provide energy security, and we need help. We're not looking to have our cake and eat it too, but we are looking for help to compete."

For the oil and gas industry, the subsidies have been built into the permanent tax code through the years. That's not the situation for biodiesel and other renewable energy advocates, who regularly must go hat in hand to Congress when their credits expire, seeking to get them reinstated. "The oil industry enjoys a lot of their tax credits and incentive programs in the baseline tax code," Rehagen says. "They don't have to go back." He doesn't begrudge the petroleum industry its production incentives; rather, he says, those subsidies over the past century must indicate that Congress recognizes the need to provide long-term support for an important energy resource. Our lawmakers should do likewise for other energy resources that they consider worthy.

Subsidies written into the tax code long ago should be revisited to ensure that they still are applicable, Whitehead says, particularly if they support an industry that is standing well on its own. "If an

---

27   Mark J. Perry, "Does the Oil-and-Gas Industry Still Need Tax Breaks?" *The Wall Street Journal* (November 13, 2016): https://www.wsj.com/articles/ does-the-oil-and-gas-industry-still-need-tax-breaks-1479092522.

industry is worthy of taxpayer money, then it deserves a chance to prove itself," and the biodiesel industry has been working diligently to do just that. "As soon as the economies of scale reach a point where that subsidy is not required, then we will move on without it. But we have a vision of 10 percent of the diesel market by 2020, and that's been in our mission statement and strategic plan for some time now."

The biodiesel industry has created thousands upon thousands of jobs, Rehagen points out, along with a variety of other clear benefits to our national interests and to the environment, and he wishes to keep up the good work. "The investment that we can make in our industry to grow is going to be tied directly to the amount of support and encouragement that we get from the federal government through the policies of the tax credit and the RFS. We would love to see a long-term certainty that the government is interested in growing our industry." That certainty would go far, he says, toward stimulating further investment by biodiesel producers and the institutions that back them.

The petroleum industry certainly is not the enemy, Rehagen emphasizes. "We rely on the petroleum industry. They're basically our customers," he says. Biodiesel, after all, finds its way into the fuel supply as a blend with petroleum diesel, and so, in that way, the two industries are partners. Here is how Rehagen sees that partnership:

"You only have to look at our federal government structure to realize what we believe are the most important components of an economy and of a successful nation. There is a Department of Energy. There is a Department of Agriculture. Our country has admitted and accepted the fact that it's very important to maintain a very consistent supply of energy. That's what drives business. But it's also very important to have an agricultural economy that works so that people have food.

With biodiesel, we are able to take those two interests and push them together. The agricultural industry—through the production of soybean oil, canola oil, and animal fats—is now able to introduce those agricultural products into the energy sector as well."

As Rehagen explains that partnership between our nation's agricultural and energy interests, he emphasizes anew that nobody goes without food simply because our nation's farmers are producing the raw materials for biodiesel. "We are very proud that all of the products that we use in biodiesel production are all byproducts or waste products from some other process. Soybeans are not grown for oil. They're grown for the protein that goes primarily into the animal-feed market. We're able to use the oil that's the byproduct to create value to the agricultural community and also value for the American consumers who have an alternative at the pump."

CHAPTER 8

MAKING IT HAPPEN

Henry Ford dreamed that the American farmer would produce the fuel for his Model T, which became the car of the people in the early twentieth century. Born and raised on a farm near Dearborn, Michigan, Ford was faithful to his roots and envisioned an agricultural source to power the nation's vehicles. "If we want the American farmer to be our customer," Ford stated, "we must find a way to become *his* customer."

That much is true—but then fact and fiction begin to overlap, as they often do. A persistent

Ford Model T

story maintains that oil tycoon John D. Rockefeller, the nabob of Standard Oil, sabotaged Ford's efforts by funneling millions of dollars into the Woman's Christian Temperance Union, which was pushing for passage of Prohibition. With the ratification of the Eighteenth Amendment in 1920, alcohol was banned for use by both man and machine. Rockefeller thereby purportedly managed to shift the prevailing automotive fuel to gasoline, finally putting the unwanted and highly combustible byproduct to a profitable use rather than continuing to dump it downriver.

It makes for a good story. It's also bunk. Although Rockefeller no doubt was keenly interested in advancing the interests of the petroleum industry, he supported the temperance movement well before the age of the automobile and his own rise to corporate power. He was a lifelong teetotaler, and had married an early activist in the movement. It is highly unlikely that he was feigning support of Prohibition for ulterior motives.

In any case, the Eighteenth Amendment did not ban the industrial use of alcohol. It banned only the manufacture, sale, or transportation of intoxicating liquors for use as a beverage. To survive, some distilleries switched to the production of alcohol for industrial purposes, including fuel. That didn't exactly advance the cause for gasoline—but it didn't much matter. By the time Prohibition began in 1920, gasoline was already entrenched as the fuel *du jour.* Standard Oil had no cause to fear that the motoring public would abandon gas, despite the popularity an ethanol blend would gain in the Midwest. During the ensuing years, in fact, the company began adding ethanol to gasoline to reduce engine knocking and to increase the octane level. The oil companies later switched to lead as the preferred additive.

Henry Ford nonetheless argued in favor of a homegrown fuel from the farms, and he made sure that his Model T could run on ethanol as well as gasoline or kerosene, or a combination of the three. Ford considered alcohol a "cleaner, nicer, better fuel for automobiles than gasoline." A knob on the dashboard controlled the carburation mixture so that the driver could adjust for the fuel type.

The Model T brought motoring to the masses. When he introduced it in 1908, Ford hoped it would "democratize the automobile"—and with the advent of mass production, the initial price of $850 did come down to $290 by 1925, near the end of the Model T's run. Adjusted for inflation, that means the price fell from about $23,000 in today's dollars to about $4,000. Such was the power of the assembly line to make the automobile affordable for the working class as well as the wealthy. At a time when a typical salary was about $1,300 a year, a family could buy a Model T for less than three months' earnings.

Long a luxury, the automobile had been transformed into a necessity. It was now part of the American way of life. With the lower prices, the demand soared. By the beginning of the Roaring '20s, nearly half of the cars in the country were Model T's, and, by 1927 when production ceased, 15 million had been sold. Likewise, demand soared for the fuel to fill its tank—but that was mostly gasoline, which was cheap and plentiful. The Model A (introduced in 1927 in response to competition from the up-and-coming General Motors) and other successors to the Model T were designed for gasoline only. For practical and economic reasons, Ford eventually gave up on his alternative-fuel vision.

For years, however, Ford continued to advocate for ethanol's potential and for finding new industrial markets for whatever farmers could produce. "The fuel of the future is going to come from fruit like that sumach out by the road, or from apples, weeds, sawdust—almost anything," Ford told a *New York Times* reporter in 1925. "There is fuel in every bit of vegetable matter that can be fermented. There's enough alcohol in one year's yield of an acre of potatoes to drive the machinery necessary to cultivate the fields for a hundred years."

The cultivation was not going well in those years, however. Many farms faced an economic crisis in the 1920s, and that would only worsen with the coming of the Great Depression and the Dust Bowl years. In the "Chemurgy" movement of the 1930s, Ford lent his financial and political backing to scientific research that could lead to new markets for farm products—and not just as a fuel.

The soybean particularly intrigued him. From the soy oil, scientists could extract glycerin and other products to produce shock absorbers and enamel. From the soy meal, they worked to develop plastics to make horn buttons, knobs, trim, and casings. "You will see the time," Ford predicted, "when a good many automobile parts will be grown." In addition to soybeans, Ford experimented with hemp in the manufacture of car bodies.

Ford's experimentation with agricultural sources for making car bodies and parts continued into the early 1940s, but he abandoned his initiative as auto production declined during World War II.

Henry Ford's Soybean Car

Researchers at Ford's soybean laboratory in Dearborn even developed a prototype vehicle known at the time as the "soybean car." Ford touted it as a lighter, safer vehicle, and he hoped it could help the auto industry as a hedge against steel rationing during the war. Exhibited at the Michigan State Fair in 1941, the prototype was later destroyed, and production records were lost.

Ethanol did see a boost in production during the war effort, however. The first World War likewise triggered a demand for industrial alcohol, but in both cases the low price and copious supply of post-war gasoline continued to kill the incentive to develop an ethanol market in the United States. It wasn't until the 1970s that the oil embargo and other crises affecting the availability and price of petroleum rekindled US interest in fuel from agricultural sources. It was during that decade as well that the potential of biodiesel gained some temporary attention until tensions eased, oil prices fell again, and the public and politicians seemed to forget what had happened.

What has kept petroleum on the throne, in other words, has been its cheaper cost. That was what happened in the early years of the last century, as well. Rockefeller and Big Oil did not conspire to kill Ford's vision. What mattered was the market. We had gushers in Texas. Biofuels were unable to pose a serious challenge to the less-expensive and widely available fossil fuels.

Despite his best efforts, Henry Ford failed in his time to develop inroads for fuels from the farm. He tried to make it happen, but by and large the public wasn't on board with his vision. They loved his Model T, but they would fill it up on the cheapest commodity available.

Renewing the Vision

Today, the case for biofuels, and specifically biodiesel, is many-faceted. Yes, this is an industry that can help the farmers by finding a broader market for their crops. Henry Ford certainly would have approved of that. But biodiesel also solves a variety of problems that weren't on the radar in his day, and chief among them are energy security and environmental issues.

Let's look now at the examples of three progressive leaders—one in Iowa, one in New York City, and one in Washington, DC—who have taken concerted action to make it happen. They are the 2017 winners of the Eye on Biodiesel Award from the NBB. Their accomplishments are inspiring and encouraging, and their contributions are helping to ensure that the industry will continue to grow. There are many others like them across the land—and many who might be like them, if only they hear the message loud and clear that this is what the people want.

> **As we continue to grow, it becomes increasingly important that we tell our industry's story of being an American-made, clean-burning, advanced biofuel.**

"The biodiesel industry has seen record growth over the past decade, which means the fuel is reaching more markets than ever before," Kent Engelbrecht, the chairman of the NBB, commented as the board recognized the award winners. "As we continue to grow, it becomes increasingly important that we tell our industry's story of being an American-made, clean-burning, advanced biofuel." These three men, he says, are among those who have gone above and beyond to advance the industry.

A Passion in Iowa

Terry Branstad

When Terry Branstad became the ambassador to China in 2017, he had been the longest-serving governor in United States history, serving as Iowa's chief executive for more than twenty-two nonconsecutive years since 1983. It was early in his first term that he met Xi Jinping, who at the time was a young agricultural official and chemical engineer visiting from Hebei Province to learn about US farm practices. They kept in touch over the years as China became a key US trading partner. Xi was elected as China's president in 2013, and, now on the fields of global diplomacy, the two work together again.

Terry Branstad

A state of rich, deep soil and wide expanses of farmland, Iowa has long been a national leader not only in agriculture but also in clean-energy production. Thanks in large part to the strong leadership of Branstad and others, Iowa is the top biodiesel-producing state in America. Branstad has been recognized repeatedly for his advocacy of renewable fuels, and some of the last legislation that he signed as governor included new and renewed policies supporting the biodiesel industry in Iowa. Shortly before he stepped down as governor to take on his new assignment in China, the NBB honored Branstad for his work and unwavering dedication to the industry.

The accomplishments include legislation that promoted jobs, cleaner air, competition at the pump, and lower fuel prices in the state. Iowa's biodiesel production credit was scheduled to expire, but Branstad extended it through 2024. Branstad also extended and expanded the biodiesel promotion retail tax credit through 2024, which encourages retailers to carry increasingly higher blends. In a tough budget year, he secured another year of funding for the Renewable Fuels Infrastructure Program—the state's successful biodiesel and ethanol blender-pump program. The legislature had previously ended funding for the program.

Those are progressive initiatives designed to keep Iowa at the forefront of biodiesel production, and to use to build the state's economy and protect the environment. Branstad also has helped to promote renewable fuels at the national level as well through his service on the Governors' Biofuels Coalition. "As I transition to my new role," he said, "I will continue to be a tireless and energetic supporter." His lieutenant governor, Kim Reynolds, who stepped up to become Branstad's successor, vowed to be just as passionate about the industry.

Iowa leads the nation in biodiesel, with a dozen plants that in recent years have set production records—although the expiration of the federal biodiesel tax credit was giving producers cause for concern near the end of Branstad's term. Soybean oil makes up about two-thirds of the feedstock for those plants.

New York Initiative
Costa Constantinides

New York City council member Costa Constantinides—the environmental champion whom we discussed in chapter 3—received the NBB's praise for his support of the industry, specifically for his legislation increasing the amount of biodiesel blended into heating oil, eventually reaching 20 percent in 2034.

A long-time advocate of improving air quality in the city, Constantinides has been advancing policies to promote cleaner-burning forms of renewable energy for their health benefits. He is chairman of the council's Environmental Protection Committee. Before he was elected to his council seat, he worked on environmental initiatives for several years as legislative director for former council member James Gennaro, who led the initial charge to require a 2 percent blend in heating oil.

"I represent a community where air quality is a very serious issue," Constantinides says. "We have power plants. We have the Grand Central Parkway. We have the airport surrounding our district. We know our kids are missing school. We know our seniors are ill. We know our most vulnerable citizens are under attack from greenhouse gases and pollution."

Top (L-R): Donnell Rehagan, Costa Constantinides, Kent Engelbrecht

Bottom: Costa Constantinides is honored by the NBB in 2017 as the Eye on Biodiesel award winner

He says he and other advocates of the legislation encountered resistance, particularly from the American Petroleum Institute. "This was a dollars and cents issue to them. We were going to burn millions fewer of gallons of petroleum. They spent hundreds of thousands of dollars to try to kill our bill," including the deployment of expensive lobbyists. He says they spread false stories through the press.

The opponents, he says, trotted out bad studies trying to show that the Bioheat effort somehow would increase pollution—"and of course they requoted the old attacks on biofuel, and the food-vs.-fuel argument." He even heard personal attacks on his integrity and his Greek heritage.

"There were so many ridiculous arguments," he says. "They showed up at my hearing. I asked them, 'Do you believe in climate change, and do you believe petroleum plays a role in manmade climate change?' They would not answer the question. They got into a shouting match with me about it. They're all about selling their product."

It's not a matter of cost, he says: a Bioheat blend generally is no more expensive for homeowners and often is cheaper. "We feel confident that Bioheat has been a boon for homeowners. We're allowing them to be green while keeping green in their pockets."

Bioheat is only one of the initiatives that Constantinides has endorsed, including the use of biodiesel in city equipment. "Our city fleet is at B10," he says, "and I have a bill that would require the city to use B5 when we institute citywide ferry service. We are looking at every mode of transportation and seeing how we can move the ball forward."

In addition, he says the City Council has been looking into geothermal technology, as well as solar readiness in retrofitting thousands of city-owned buildings. "We are working on wind power, and on electric vehicles" with publicly accessible charging stations. New York, he says, is a world city, and it sets the example for other metropolitan areas, particularly in the Northeast. The leadership must do what is right. Climate change is undeniable, he says. Sea levels are rising, and severe weather events could put New Yorkers at risk. "That's why we decided we were going to fight for a progressive agenda to reduce our emissions."

Why should the citizens care? "You're going to be helping to protect your city in the long term against climate change," Constantinides says. "You're going to reduce your costs on home heating. You're going to see an economy that is more diversified and less dependent on Wall Street by creating green jobs and creating new industries and new tax bases. You're going to strengthen your city, make it more resilient, more sustainable."

Education must be part of the mission, says Constantinides. The public and the policymakers need to understand what is truly at stake. "We have to stand up to big interests. It's not going to be simple. The American Petroleum Institute is millions of dollars. We're going to have to stand firm and say that climate change is real and hold our elected officials accountable for doing something about climate change and taking real action, not just lip service. If someone says they're not sure about it, that means they're not fighting for the future of our city or our state or our country. We have to know the language, know what's actually happening, and be able to push back."

Keeping it Clean in DC
Ron Flowers

An educator at heart, Ron Flowers is both a teacher and a doer. He has spent a significant part of his career spreading the word about biodiesel and working diligently to get it into the fuel supply. From 2010 until he retired in 2017, Flowers was the executive director of the Greater Washington, DC Region Clean Cities Coalition.

The biodiesel industry knows him as a voice of wisdom with nearly half a century of service in the public and private sectors. He played a key role in guiding the district government, the Smithsonian Institution, American University, and many others as they began using

Ron Flowers

biodiesel in their fleets. Many of those who converted have themselves become champions for the industry.

On the eve of his retirement, the NBB commended Flowers on his enduring influence on both the industry and on the health and welfare of society. "Retirement," however, is relative. This is a man who is determined, as he puts it, "to stay in the race." He is developing a fleet-consulting business in the Washington region. He sees it as a forum for continuing to use his experience and knowledge to teach about alternative fuels, providing solutions along with the equipment. "I have a compelling interest to demonstrate and articulate that the use of bio-based products as well as biodiesel is extremely important to the health and welfare of our country and of the world. I'll continue to do that in some form or other."

Flowers certainly has, in some form or other, been involved in education since his graduation from Chicago State University with a degree in occupational education. His roots also are in fleet operations, which was long the focus of his family in Chicago. In both cities, Flowers served in university positions (he came to Washington as vice president for students at the University of the District of Columbia) as well as school and government transportation positions. As an experienced educator with fleet-management experience, he was a natural to eventually take the helm of the region's Clean Cities Coalition, a Department of Energy incentive program encouraging less dependence on petroleum.

His involvement with alternative fuels dates to about 1990 when he was transportation director for the city schools. The school system at the time received a grant from the Department of Energy to purchase four transit buses that operated on compressed natural gas (CNG), and also got funding to install a CNG fueling station. Flowers had been concerned that dangerous fumes were surrounding students, including children with special needs, during their daily transportation. In 1993, about the time that the Clean Cities Coalition was established, Flowers became chief of the district government's fleet operations, and began working for alternative fuels funding for the city. Early efforts involved CNG and ethanol and some electric. Around 2005, the city began looking seriously at biodiesel, as well.

In the years since, the district government has reached the point where it is using about 1.3 million gallons of biodiesel fuel a year throughout the city fleets, including heavy equipment, trash and snow removal trucks, and some fire and police vehicles. The Smithsonian and the National Zoo are also using biodiesel.

In the beginning, he says, the initiatives encountered some resistance, but educational efforts largely were able to overcome that. Some early users of poor-quality biodiesel had experienced problems, he says, but as the industry has policed itself and dramatically improved the product, those issues have faded. "We had to overcome that baggage," he says. To reassure users, the district initially set up a pilot project, setting aside ten trash trucks to run on biodiesel to demonstrate that they would operate effectively. In addition, those who serviced the vehicles needed training to clean filters and tanks regularly as the biodiesel begins to clean out engine sediments. They also needed training on the proper winter blends and additives and other specifics of managing a fleet on biodiesel.

Flowers speaks proudly of other biodiesel-fleet programs in the region. As examples, he points out that Arlington County, Virginia, was an early leader, and that a number of campuses have gone biodiesel with their shuttle services for students and staff. American University wanted to use biodiesel in its shuttles, but the campus lacked the space to install a fueling station. The university tried to procure biodiesel through a public station, but often the fuel was unavailable and it became expensive to drive to the station and back. The university worked with city and fire department officials to arrange for a mobile biodiesel fueling tank on campus. The university leaders are committed, Flowers says, to doing whatever they can to promote a green and energy-efficient campus environment. Likewise, Georgetown University has a green fleet program and uses biodiesel in its shuttle buses. At George Washington University, the shuttles operated by International Limousine Service, based in the city, are fueled with biodiesel. The company's entire bus fleet operates on a B20 blend.

More such programs are developing in the region as the word spreads. Flowers has not been specifically involved in the Bioheat movement, but it has certainly caught his attention. He has conferred with Constantinides about New York City's initiatives and relayed the information to his Clean City successors for their consideration.

Awareness is growing, Flowers says. He sees the young people of today as active advocates for a cleaner and healthier environment, and "that's in part because of the efforts that we've been making over the years in raising the concerns." Biodiesel is a cost-efficient and easy way to help to improve the environment, and therefore it "takes away a lot of the excuses that some people will use for not getting into some form of alternative fuel." His personal motivation, he says, is a better world for his two grandchildren.

"I believe in people working together," he says. "We don't have to dwell so much on what the problem is. We need to dwell more on what the solutions are. We've got to encourage people to spend more time on the *can-do* side of it."

The petroleum industry, despite its concerns about the market and the bottom line, can be partners in the green movement, Flowers says. "A number of the companies are looking at alternative fuels as they look at long-term development, and some of them are starting to embrace the fact that alternative fuels are not going away. Biodiesel is not going away."

"The Need of Knowing"

Clearly, individuals who speak up and take action are most likely to prevail. That is the nature of politics. It is also the nature of our democratic society, however, that power lies collectively with the people, who can wield their political influence to foster big changes in our society. They can exert the sort of pressure that gets results. Once they recognize a cause, they can get behind it and exhort their leaders to do the same.

The biodiesel industry today faces challenges that could inhibit its ability to produce more to meet a growing demand.

> *It is time for the people to speak loudly on behalf of this industry that offers nothing but huge benefits and badly needed solutions. There is no downside.*

We all must do our part. Get involved. Talk to your elected officials. Contact the legislators who represent you at every level of government, and demand policies that support the biodiesel industry. Tell your heating-oil and diesel dealers why you want higher blends in the fuel supply—and explain why. Talk to your bus driver, your bus driver's boss, your school officials, and to anyone responsible for any fleet of vehicles that could be taking better advantage of biodiesel's many benefits. Whenever you hear any falsehoods or distortions, set those people straight. You know better now. You understand the full story, so what are you going to do about it? If you are convinced that we should be using more American biodiesel and reducing the carbon footprint, how will you help to make it happen?

We can accomplish much by working together to meet our energy needs and to take care of our planet. Complacency can kill us, quite

> *We can accomplish much by working together to meet our energy needs and to take care of our planet. Complacency can kill us, quite literally.*

literally. We need change, and it starts with understanding and acknowledging that we need it. Ron Flowers, the lifelong educator who led so many in the Washington, DC, region to make good decisions, said it best as he accepted his accolades for his advocacy of biodiesel: "Those who do not know," he said, "know not the need of knowing. It is important that each of us help people to understand that biodiesel plays a major role in the quality of life."

CONCLUSION

WHERE WE'RE GOING ...

Where we're going, we will need *plenty* of roads. Sorry, Doc, but we won't be giving up on highways anytime soon, even as we go "back to the future" with a high-tech fuel rooted in our agricultural heritage. We also will have plenty of trucks, cars, tractors, planes, ships, barges, and cranes equipped with modern versions of the mechanical marvel conceived more than a century ago by Rudolf Diesel.

The big difference is that those engines will be running on biodiesel rather than solely on fossil fuels. And in the world to come, we will still need to fuel the fires that keep us warm in our homes and offices—and Bioheat will play an increasingly important role.

This is the dream of visionaries from many walks of life, some of whom have shared their views in these pages. The vision is alive and well, but the challenges are urgent. If we work diligently together, we can beat the clock. If we fail to get on board, it's only a matter of time.

Biodiesel offers the prospect of a cleaner and safer planet. Fossil fuels have played a crucial role for generations, but they come with limitations. They have an expiration date. We must strive for a day when we can power the world without fear of running out of energy, and the renewable nature of biodiesel advances that goal.

Society's many pressing issues can be resolved only in the spirit of partnership. To get to where we need to go, none of us can stand alone. The best minds must collaborate for our mutual benefit. The biodiesel industry must join forces with the petroleum industry, its primary customer. The oil refiners, after all, provide the fuel for the optimal biodiesel blends. The future depends on the cooperation of the leaders of diverse industries. We need the insights of the fuel makers and the fuel dealers, the entrepreneurs and the environmentalists, the farmers and the truckers—and the politicians who represent the people.

In the people lies the ultimate power, and it is high time to step forward and spread the word about biodiesel and clear up any misconceptions. We need to rally behind the truth, as borne out by the facts.

Here is a sampling of those facts that these chapters have examined:

- Biodiesel is a clean, renewable, and efficient fuel that we can use today and tomorrow to power our world. It is helping us to conserve our limited resources as we take a big step toward energy security.

- Biodiesel can be added in various percentages to petroleum diesel to significantly reduce pollutants and greenhouse gases, a critical advantage in today's world. It is a safer fuel to produce, transport, and use.

- Existing diesel engines and heating-oil systems run just fine on biodiesel blends. The equipment requires no retrofitting. Those who have switched to biodiesel are pleased with its performance.

- The biodiesel industry puts tens of thousands of Americans to work in good manufacturing and related jobs. It could support many more families and further boost local economies if the government adopts policies to encourage our domestic production rather than stifle it.

- Although biodiesel is synthesized from vegetable oils and animal fats, the industry does not take away food from a hungry world. In fact, it encourages farmers to produce more. The industry's growth depends on the availability of agricultural byproducts that otherwise would go to waste. More food on the plate means more raw materials to make biodiesel. The industry gives value to those leftovers by converting them into an exciting source of energy.

Our leaders need to hear and heed this message: America should be producing much more of this remarkable fuel. Our nation needs strong federal policies and incentives to help build the biodiesel industry at home instead of encouraging imports. The government must fairly subsidize the industry so that it stands a chance against established interests on a level playing field. Each state should promote biodiesel production through mandates and tax incentives. Local leaders of government, education, and industry should do their part as well. School districts, for example, can insist on a fuel blend in their buses that will protect the health of our children. Municipalities can power their fleets and heat their buildings with biodiesel blends. It has never been true that you can't fight City Hall. In fact, City Hall can do much of the fighting.

The American people have been abundantly clear about what they want our leaders to support. The people want policies that build the domestic economy and create jobs. They want policies that support agriculture. They want sensible policies that address our pressing concerns about the environment and energy independence. At every turn, a thriving biodiesel industry makes sense. Our challenge is to get more of our political leaders and policymakers to recognize what their constituents have been demanding. The time to act is now. Our planet cannot wait.